On the Tribology of Lubricating Greases

TRIBOLOGIE

SCHMIERUNG, REIBUNG, VERSCHLEIß

Herausgegeben von Dr. Manfred Jungk

Die Tribologie ist ein interdisziplinäres Fachgebiet, mit Schwerpunkten aus den Bereichen Maschinenbau, Chemie, Physik und Werkstoffwissenschaften. Entsprechend vielfältig sind die Forschungsthemen und Anwendungen.

Die Reihe *Tribologie – Schmierung, Reibung, Verschleiß* behandelt sowohl Grundlagen des Themengebietes für Anwender:innen, Wissenschaftler:innen und Studierende als auch moderne Trends wie Nachhaltigkeit, tribologische Aspekte der Industrie 4.0 und Herausforderungen durch die Elektromobilität.

Erik Kuhn

On the Tribology of Lubricating Greases

An energetic approach to post-modern tribology

expert

Bibliografische Information der Deutschen Nationalbibliothek
Die Deutsche Nationalbibliothek verzeichnet diese Publikation in der Deutschen Nationalbibliografie; detaillierte bibliografische Daten sind im Internet über http://dnb.dnb.de abrufbar.

DOI: https://doi.org/10.24053/9783381141722

– Ein Unternehmen der Narr Francke Attempto Verlag GmbH + Co. KG
Dischingerweg 5 · D-72070 Tübingen

Internet: www.expertverlag.de
eMail: info@verlag.expert

Druck: Elanders Waiblingen GmbH

ISSN 2701-603X
ISBN 978-3-381-14171-5 (Print)
ISBN 978-3-381-14172-2 (ePDF)
ISBN 978-3-381-14173-9 (ePub)

MIX
Papier | Fördert gute Waldnutzung
FSC® C016368
www.fsc.org

Dedicated to my wife Kerstin

and to our children,

Josepha and Leonhard,

in gratitude for the many, many conversations.

Preface

Grease has been used as a lubricant for many centuries, proving to be a simple, efficient and cost-effective lubrication method. Grease is also a crucial component in a great number of machine elements and mechanisms, which can be optimized and tuned to very different operating conditions.

Grease has been the object of continuous studies, concerning materials (thickener, base oil, additives and nano-particles ...), physical, chemical and rheological properties, lubricant film build-up and starvation characteristics, friction behaviour, and wear minimization. In more recent years, the energetic behaviour of greases has been at the top of the research and development studies on greases, since power loss minimization is a priority in machines and mechanisms.

This book provides a clear overview of the state-of-the-art of greases, as well as a fresh and innovative approach to the energetic behaviour of greases.

Several non-usual concepts are presented, developed, and applied to lubricating greases: entropy, self-organization, energy expenditure, energy activation, and wear of greases. The influence of grease formulation and operating parameters on these concepts, and their relevance in the energetic performance of machine elements and mechanisms, is analysed in detail. New rheological and mechanical tests are also proposed, to evaluate the energy expenditure and wear of lubricating greases. These new approaches to grease lubrication, and the new concepts and tests proposed, clearly represent a postmodern analysis of grease tribology.

Jorge H.O. Seabra
University of Porto
2025

Preface

Lubrication has been practiced for thousands of years, dating back to the earliest days of human civilization. Ancient Egyptian civilizations used oil or water to reduce the energy required to move heavy stone blocks and large statues. This is evidenced by the well-known wall painting from 1880 B.C. in the tomb of Djehutihotep. Since then, humans have used bio-based materials of animal or vegetable origin as lubricants. It was only in the mid-19th century that petroleum-based lubricants came into use. With the growing demand for lubricants from the emerging automotive industry at the beginning of the 20th century, various industrial manufacturing processes were developed to produce mineral oils. Later, coinciding with the Second World War and the increasing demand for lubricants in the aviation and aerospace industries, lubricant technology advanced significantly with the development of both additives and synthetic lubricants. During this time, numerous R&D studies in the lubricant field have been conducted, allowing us to understand in great detail the friction and wear behavior of these mineral and synthetic lubricants, leading to better productivity, performance reliability, and energy efficiency.

The use of lubricants in industrial activity is essential; around 23% of global primary energy losses are due to friction. However, the increasing demand for both mineral and synthetic lubricants is causing serious environmental problems. Therefore, it is worth noting that today we find ourselves in the paradox of meeting our lubrication needs with environmentally acceptable lubricants (EAL), as our ancestors did 4000 years ago, but with unprecedented technological development that has been achieved today. Thus, it is necessary to develop scientific and technical knowledge to obtain bio-based lubricants that are as technologically advanced as those made from mineral and synthetic oils, to replace them while maintaining the current efficiency of lubrication processes.

In the specific case of lubricating greases, these are viscoelastic lubricants whose rheological behavior and microstructural characteristics significantly affect the efficiency of the friction process and the energy consumption involved. However, the mechanism by which shear-induced structural degradation contributes to energy expenditure when lubricating grease is subjected to mechanical stress during the friction process is still poorly understood. Therefore, replacing traditional thickeners, such as metallic soaps, with other bio-based thickeners poses a significant challenge for the scientific community and lubricating grease manufacturers.

This book, *Tribology of Lubricating Greases: An Energetic Approach to Post-Modern Tribology,* is presented as an innovative and essential work for those interested in the Green Tribology of biogenic greases. This book offers a novel approach by analyzing Tribology from an energetic perspective, a crucial aspect for understanding how the structural degradation of biogenic greases affects energy consumption during lubrication processes. The content of this book has been carefully structured to guide the reader from the basic concepts of Tribology to the more innovative concept of lubricating "grease wear ". To this end, Prof. Kuhn develops the concept of Entropy and self-organization in Postmodern Tribology in a practical and accessible way, with an interesting chapter dedicated to the Rheology of lubricating greases, with special attention to biogenic greases. With this structure, Prof. Kuhn clearly shows how the thixotropic behavior of a lubricating grease, influenced by the structure of the thickener used in its formulation, affects the efficiency of the friction process. He encompasses this fact in his innovative "grease wear " concept, suggesting a relationship between shear-induced structural degradation and the energy expended during the friction process.

Additionally, the book includes experimental studies and theoretical models that provide a deep understanding of how biogenic greases behave under different stress conditions and how they can self-organize to improve their performance. These kinds of studies are essential for those seeking to develop lubricating grease formulations that are both efficient and environmentally acceptable. In this sense, I would like to mention promising advances made at the University of Huelva and the collaboration with Prof. Kuhn, which demonstrate the high thickening capacity of both cellulose nanofibers and nanocrystals as biothickeners in vegetable oils. These eco-friendly nanocellulose-based lubricants have been made possible by a sustainable methanol-based solvent exchange method. With only 2.8 wt.%, they provided structural skeletons constituted by entanglements of individual fibrils and dense fibril arrangements, leading to a biogenic grease with an NLGI 2 grade. Tribological analysis concluded that these fully biogenic lubricating greases behave similarly to Li-grease, with promising anti-wear properties and no starvation at high temperatures.

In conclusion, *Tribology of Lubricating Greases* is a valuable contribution for researchers, engineers, and professionals in the field of Tribology, offering a comprehensive exploration of the energetic description of the friction process with biogenic lubricating greases.

I would like to extend my heartfelt gratitude to my friend, Prof. Kuhn, for his exceptional dedication and invaluable contributions to advancing our understanding of the energetic processes involved in grease lubrication. It has been a true honor to introduce this significant contribution to the field of Green Tribology literature.

Miguel Ángel Delgado Canto
University of Huelva, Spain
2025

Contents

1 Author's Preface

Finding answers to questions that may have been asked for ages is a great fortune and a deep satisfaction. To pose questions that open up a path filled with the hope of entirely new and unexpected answers is a state of bliss.

Tribology, as the *science of friction*—or, as A. KNAPPWOST [1] describes it, the study of frictional processes—deals with a fundamental natural phenomenon.

To engage with friction means acknowledging the irreversible character of all natural processes and inquiring into the driving forces behind observable mechanisms. This often experimental observation of mechanisms has a long history, as frictional effects are encountered by humans in countless situations—both in everyday life and in artificial configurations. For this reason, the conscious attempt to influence and modify frictional effects in the observer's interest has an equally long historical development, culminating today in an almost unmanageable number of tribological research projects.

A genuinely new and fundamental development in tribology has emerged through the observation and analysis of systems striving to maintain stable process states through energy dissipation—or, when this is not possible, to form new structures from a state of *instability*.

After, among others, I. PRIGOGINE [2] paved the way for the investigation of open systems that exist far from equilibrium in order to understand processes initiated from within the system, tribology has begun to follow this path.

Fifty years ago, A. KNAPPWOST [1] already pointed out the mistake of attributing the early development of tribology solely to mechanics and of not linking it more intensively with the fundamental laws of thermodynamics.

Instability as the core of natural processes, as the source of new structure formation, as the prerequisite for new emergence, led to the conceptual framework of postmodern natural science. I would like to take up this proposal by J.C. SCHMIDT [3] and refer to this new path in tribological research as *postmodern tribology*.

This book presents works that take up this new path and places grease tribology in an entirely new perspective. I thank the publisher for their willingness to publish the proposed expansions.

Practical considerations and fundamental information on the special lubricant *grease* follow.

My intention is to use the example of grease tribology to formulate some fundamental questions—questions about the natural driving forces that resonate in the observed phenomena and whose recognition will remain a major task in the future.

Erik Kuhn
Hamburg

2 Definition and System Considerations

2.1 Definitions

2.1.1 General Terms

In this chapter, some important definitions are provided as a preamble to the subsequent discussions of various studies. The detailed treatment and justification of the selected definitions will be provided later in the corresponding sections of the book.

The tribological terminology connects the perspectives of very different researchers in describing the phenomena under investigation. However, it is also subject to development, reflecting the expansion of knowledge through changed definitions. My work has repeatedly led to viewpoints and definitions that deviate from conventional or standardized descriptions. These will be presented in this book and contrasted with traditionally established definitions.

Definition 2.1 *(Tribology) The scientific discipline of tribology investigates the endeavor of stressed and thermodynamically disordered tribological systems to regain a stable situation (a stationary non-equilibrium).*

Another, traditional definition is described in Arbeitsblatt 7 of the Gesellschaft für Tribologie [4].

Definition 2.2 *(Tribology-GfT) Tribology is the science and technology of interacting surfaces in relative motion. It encompasses the entire field of friction and wear, including lubrication, and includes corresponding interfacial interactions between solids as well as between solids and liquids or gases.*

From a perspective particularly interesting for this book, B.I. Kosjetzki [5] describes tribology.

Definition 2.3 *(Tribology-Kosjetzki) The theoretical basis of tribology is a structural-energetic conception based on the modern insights of theoretical physics in the field of self-organization and the formation of new phases of dissipative structures.*

The abstraction of any friction pair for the examination of the involved elements and their interactions presents the simplified tribological system according to [4].

Definition 2.4 *(Tribological System (Extract)) The components and materials directly involved in wear are referred to as elements of the tribosystem. Together with their tribologically important properties and interactions, they characterize the structure of the tribosystem...*

This simplified tribological system has 4 system elements: friction body 1, friction body 2, the intermediate substance, and the surrounding medium. Elaborations on the system analysis of tribological processes can be found, among others, in works by CZICHOS [6], [7], FLEISCHER[8], [9], SALOMON [10], BAUER [11].

2.1.2 Tribological Contact

For friction and wear analysis, contact geometric investigations are essential in many cases. It is defined as follows:

Definition 2.5 *(Tribological Contact) Tribological contact describes the geometric arrangement of friction bodies with each other, where interactions occur in terms of the tribological process [12].*

2.1.3 Friction

Definition 2.6 *(Friction) Friction is the input of mechanical energy into a tribological system [13]*

The investigations I have conducted in this book lead to this definition.

Now, let's elucidate the more detailed terms relevant against the background of the investigation issue. Principally, a distinction is made between *friction states* and *types of friction.* For the classification into friction states, according to DIN 50323, the state of aggregation of the involved material areas is the criterion. This results in solid friction, liquid friction, and gas friction.

Definition 2.7 *(Solid Friction) Friction state in which the stressed material areas exhibit solid properties.*

Definition 2.8 *(Liquid Friction) Friction state in which the stressed material areas exhibit liquid properties.*

In tribology, the use of the term "liquid friction" is almost exclusively associated with describing friction in a lubricant. Strictly speaking, lubricants do not fulfill all the criteria of the definition of a liquid. This is immediately evident in the case of lubricating greases, since there is friction in the lubricant, but this term doesn't adequately correspond to solid friction in the hierarchy of a terminological listing - this would then be liquid friction. The state of gas friction can be deduced from the definitions listed and is not further described.

It is possible for all friction states to coexist simultaneously.

Definition 2.9 *(Mixed Friction) Mixed friction refers to the simultaneous, adjacent occurrence of different friction states. Mixed friction is not a friction state.*

The interpretation of the definition of mixed friction provided by [14] suggests that the solid friction component in mixed friction occurs locally not alternatively to liquid friction but rather alongside or on top of it. This conceptual model therefore does not necessarily assume a breakdown of the lubricating film. In the argumentation, this definition is based, among other things, on the circumstance demonstrated by [15] that a mechanical penetration of the lubricating film is impossible when considering the pressure-viscosity effect (there exists a finite lubricating film thickness). Mixed friction with a closed lubricating film is the most common friction situation encountered in lubricated pairs and is associated with the *normal* operating condition.

Definition 2.10 *(Friction Form) The friction form indicates how many friction states are present in the considered friction contact. Three types of friction forms are distinguished: Unal friction, Dual friction, and Trial friction.*

I cannot help but mention an interesting representation of friction presented by HANS UMSTÄTTER in 1948 [16]. According to him, „friction is characterized by the difference between kinetic and potential energy. This increase in kinetic energy content through the dissipation of viscoelastic stresses perhaps best characterizes the irreversibility of friction processes.“

2.1.4 Wear

Explanations, comparisons, in-depth research, and experimental investigations on wear will be conducted in later chapters. In connection with the given definition for friction (2.6), it is stated [17],[18]

Definition 2.11 *(Wear) Wear is the dissipation of energy supplied in the friction process while simultaneously producing entropy and it encompasses all elements of a tribological system.*

The definition of wear according to the former DIN 50323-2 defines wear as progressive material loss from the surface of a solid body.

For the inclusion of the intermediate substance (a lubricant) in a wear description, the introduction of a second wear state is proposed. I made considerations for a second wear state at a very early stage [19], [20].

Definition 2.12 *(Solid Wear) Solid wear is the dissipation of supplied friction energy in stressed material areas of the involved solid bodies.*

Definition 2.13 *(Liquid Wear) Liquid wear is the dissipation of supplied friction energy in stressed material areas with liquid properties.*

2.1.5 Lubricating Grease

In accordance with the main subject of investigation of this work, lubricating grease and lubricating grease structure shall be defined.

Definition 2.14 *(Lubricating Grease) Lubricating grease is a colloidal dispersed system. It consists of a base oil and a solid substance. Lubricating grease have distinct viscoelastic properties.*

Definition 2.15 *(Lubricating Grease Structure) The lubricating grease structure is characterized by the distribution and geometric shape of the solid substance in its interactions.*

2.1.6 System Considerations

Traditional works and presentations on the *Tribological System* come from [21], [22], [9], among others. The investigations of H. CZICHOS and J. MØLGAARD have particularly influenced the conceptualization of systems in tribology and continue to be the basis for many research efforts to describe tribological contact systems. They describe the function and structure (see Figure (1)) of a tribological system and divide the entire tribological process into consideration levels.

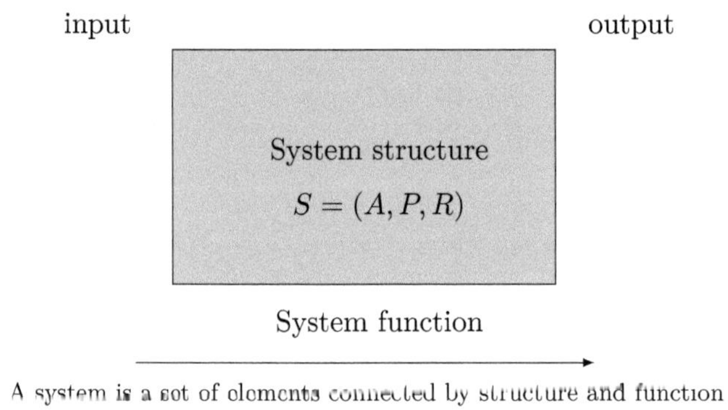

Figure 1: General system representation according to [6], [22]

In Fig.(1), S represents the system structure, A represents the system elements $\{a_1, a_2, ...a_n\}$, P represents the properties of the elements $P = \{P_1(a_1), P_2(a_2), ...P_m(a_n)\}$, and R represents the interactions of the elements $R = \{R_1(a_1...a_k), ...R_l(a_1...a_k)\}$. Here, n = number of elements, m = number of properties of an element, l = number of interactions between k elements $(2 < k < n)$.

Offering an external perspective on the subject of consideration, H. CZICHOS and J. MØLGAARD introduce a black-box representation. This serves "to describe the function of mechanical systems with regard to the transmission and conversion of the quantity that the system is supposed to process."

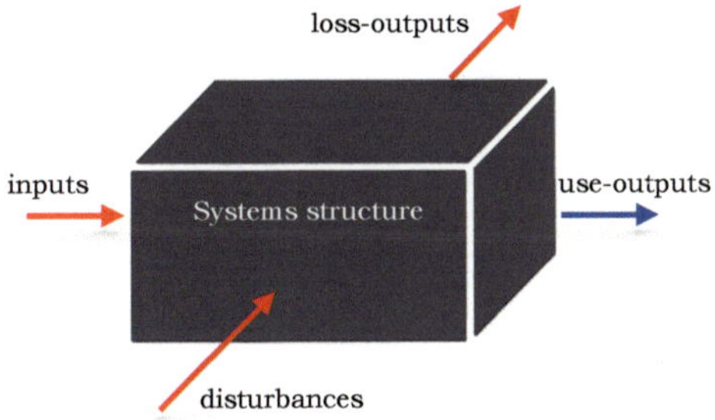

Figure 2: Black-box view of a mechanical system designed according to [6]

The authors examine three functional levels: a level of functional performance, a level of mechanical work, and a level of material. They interconnect the different properties.

Their endeavor is to investigate properties related to "the loss of useful work due to friction and vibration, properties related to entropy transport, and properties related to the displacement and transformation of materials, i.e., wear processes"[21].

Their conception of wear is also fundamental. For them, the basic characteristic of wear is the removal of material. Thus, they emphasize the application of the term *wear* to all processes that lead to abrasion.

Of particular note in the detailed analysis of each level is the consideration of entropy transport and entropy production. For instance, the authors describe the thermal level (see Figure (3)) as the conversion of mechanical work into heat accompanied by phenomena of heat conduction and thermal radiation. This involves considering an entropy production rate and entropy transport.

It is noteworthy that [21] develop their considerations against the background of the friction contact of machine elements.

Overall, 8 levels are analyzed in detail, describing the tribological process. [6] states:

- The structural description is an internal description, i.e., an attempt to understand the behavior of the system from the elements, their properties, and their mutual interactions.

- In contrast, the functional description is an external description, as it characterizes the behavior of the system in its interaction with the outside world.

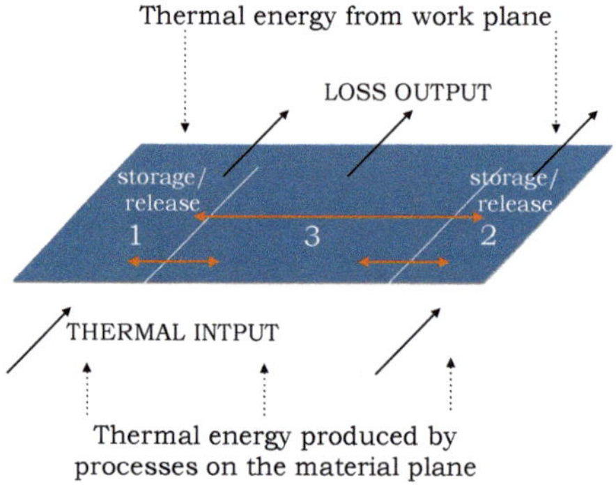

Figure 3: Thermal plane with thermal transaction designed according to [6] with 1, 2 the first and second machine element and 3 the interfacial volume

The starting point is the general structure of the system under investigation in Figure (4). BAUER [11] claims to conduct a *holistic system-analytical examination* considering all aspects of the tribosystem, expanding on the presentation in [23].

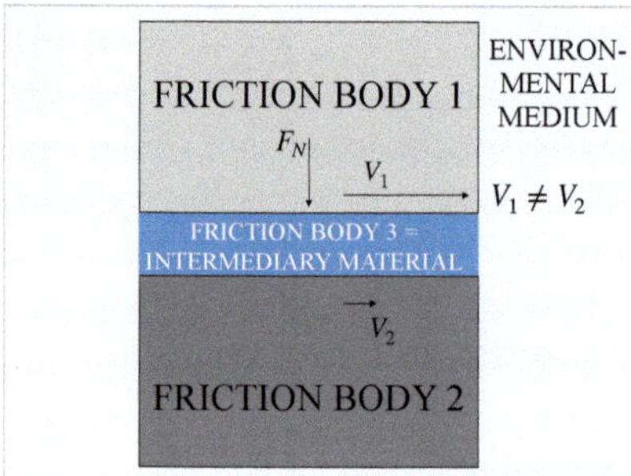

Figure 4: The general tribological system, similar in [11], [23]

He analyzes in detail (see Fig.(5)) the *stress spectrum*, the *disturbances*, and *reference variables*. Similarly, *utility variables* are extensively defined and examined, *loss variables* are named and classified. A presentation is made regarding the relations and effects of the variables among themselves.

Despite this very intense discussion of describing the tribological contact as a *tribological system*, attention shall now be focused on the role of the occurrence of instabilities as explained in this book, and the tribological system shall be reflected upon accordingly.

An important note is found in [22]: "To further simplify, the system considerations at this stage are limited to a steady state, i.e., a state of dynamic equilibrium. In this case,

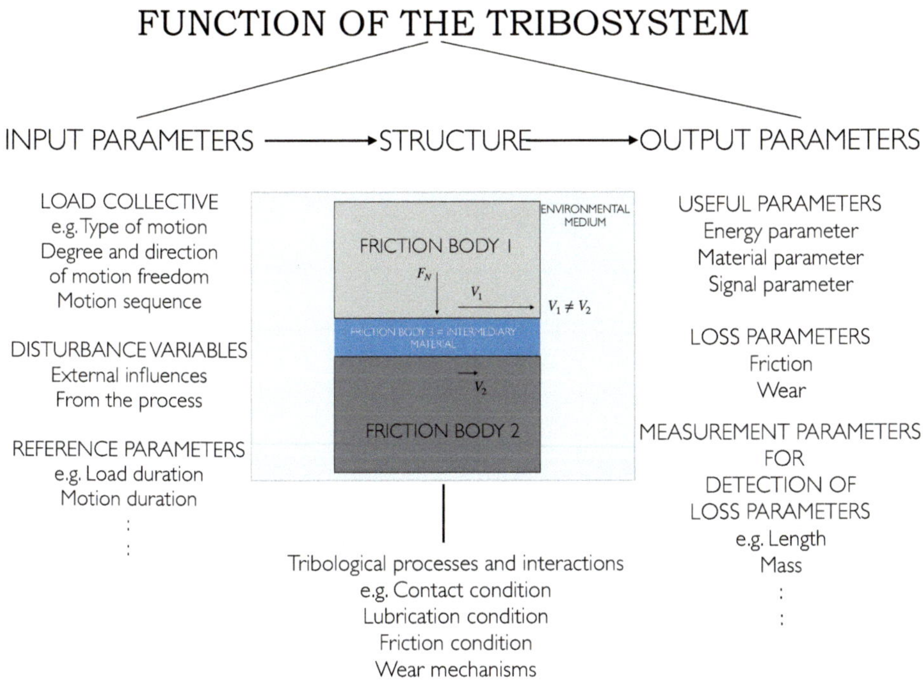

Figure 5: F. Bauer extends the consideration of the tribological system, simplified representation according to [11]

it is assumed that the system parameters are constant on average over time." A slightly different perspective emerges when the focus is shifted more towards the structure of the system, and the examination is conducted not from the user's perspective but from the system's perspective.

This means that input and output variables are analyzed but not evaluated. Categories such as *utility* variables, *loss* variables, or *disturbance* variables do not arise.

The understanding of wear at [22] et al. is expanded by the conception of wear presented here and is also reflected in system considerations. The consideration of output variables in system representations as effects suggests a meaningful examination of the system's state (see Figure (6)).

The entire process is driven by the states *unstable* and *stable.*

Driving forces are the gradients that occur. In all ongoing processes, the system's endeavor is to remain in a stable position or to regain it.

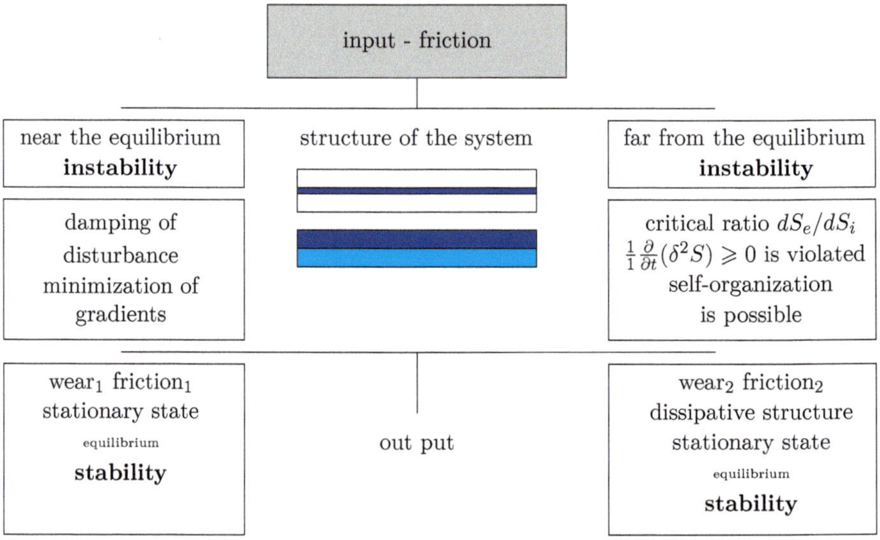

Figure 6: The tribological system, including system states

It shall be further elaborated:
If one follows the concepts of the friction states, and describes, for example, mixed friction as the presence of solid and liquid friction components that occur simultaneously but not alternately, according to the definition (2.9), the need arises to formulate a *subsystem* (Fig. (7)). It is the smallest unit of consideration in the description of the general (macroscopic) tribological system.

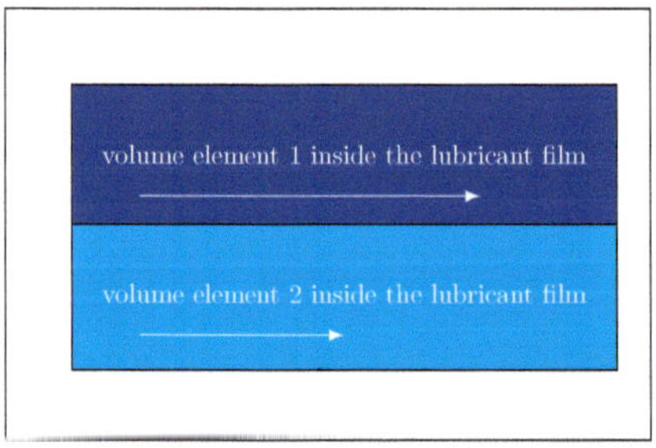

Figure 7: Structure of a tribological subsystem

Definition 2.16 *(Tribological Subsystem) The tribological subsystem consists of two defined friction bodies in relative motion that form a tribological contact (according to Def. 2.5). It is situated within a general tribological system and receives its loading from it. It becomes necessary in the simultaneous presence of multiple friction states.*

3 Introduction to Instability and Postmodern Tribology

In traditional natural science, and particularly in classical tribology, concepts such as *experimental reproducibility* and *mathematical predictability* play a central role. When regularities are found in recorded data during experimental investigations, it becomes possible to model these observations, potentially formulating laws. This procedure is based on the stable behavior of the observed phenomena. Fluctuations, irregularities, or so-called outliers are generally considered disturbances, experimental errors, or other deficiencies and are eliminated.

Undoubtedly, an unspoken assumption of stability has been the foundation of natural sciences for centuries. This assumption underlies many investigations and, in tribology, the majority of works to this day.

In a comprehensive work, J.C. SCHMIDT[1] writes about the historical development of the understanding of *instability*, "What appeared as unstable was assumed to be merely a disturbance of the stable"[3]. It was MAXWELL and POINCARÉ who introduced the unstable side of nature to a new scientific evaluation. However, even for them, instabilities did not yet represent the source of self-organization and structure formation (see [3]). Since the mid-20th century, there have been increasing publications that place the occurrence of *instabilities* and thus the recognition of unstable states as *the other side of nature* (J.C. SCHMIDT) at the center of focus.

I. PRIGOGINE writes with a focus on self-organization, emphasizing how fundamental the understanding of stability and *instability* is to describe the irreversible character of all processes [2] . He demonstrates convincingly how the occurrence of *instabilities* makes the development of new structures possible in the first place.

SCHMIDT expresses it even more succinctly: "Not stability, ..., but *instability* is considered the fundamental character of nature "[3].

It can be inferred that, for tribological research, phenomena such as process instability, descriptions of irreversibility, and nature's ability to self-organize should increasingly come into focus.

The philosopher and physicist J.C. SCHMIDT reclassifies the natural phenomena of *instability* and irreversibility within the philosophy of science and situates them within a long epistemological development.

In the same context, one finds a terminological redefinition. [3] writes, "As long as the mathematical sciences recognize and articulate the *unstable* in nature, technology, and society, they can be described as *postmodern*".

"The term stands for a very sober description of changes in the fabric of the sciences, their objects, methods, and content—from mathematics to physics, chemistry, biology,

[1]The original quotes of J.C. SCHMIDT in this chapter are displayed in German at the end of this book.

computer science, and engineering to the social sciences."

When we look at the development of tribology, it is all too evident that an assumption of stability characterizes the overwhelming majority of works. Experimental work has always relied on the guideline of achieving the highest possible reproducibility of results. Mean value considerations are deemed acceptable if fluctuations remain within a predetermined framework. Many scientific journals require indications of reproducibility, disturbances are often eliminated, and not included in further considerations.

Since this approach has produced and continues to produce considerable success, particularly driving unprecedented development in technical applications, the aforementioned assumption of stability has evolved into a general research paradigm in tribology. Tribological inquiries have focused on understanding the mechanisms in play and how they can be influenced. This is understandable since tribological research often occurs against the backdrop of specific friction pairs, such as rolling bearings or sliding bearings.

A shift, indeed a change in the described paradigm, can occur if two requirements are posed and met [17]:

1. Shift the perspective from the observer's side to the side of the system under investigation.

2. Change the questions from those about the ongoing mechanisms to those about the driving forces of the tribological processes.

By doing so, the specific conditions of a chosen friction pair can be left behind. The examined tribological system eludes our evaluation and sets its own standards.

The application of the so-called entropy concept, the consideration of instability triggered by fluctuations, and the consistent pursuit of process stability as an inherent driving force, as a self-organizing structure, make the designation **postmodern tribology** possible. In doing so, it complements and expands the approaches developed and applied up to that point.

Not least, it can be seen that the cause-effect-chain in the traditional representation (Figure 8) can be altered [17].

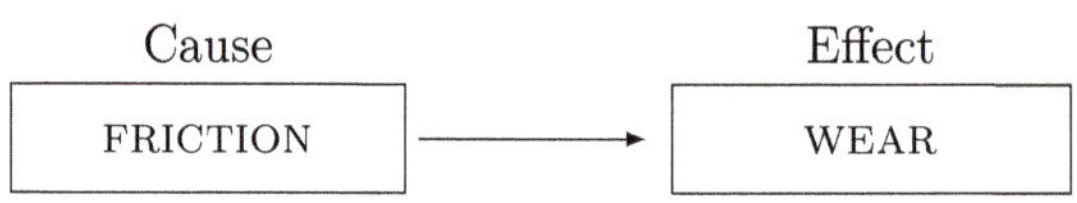

Figure 8: Traditional cause-effect-chain

With Figure 9, a differentiated view is presented, where the occurring wear, regardless of its form, leads back to stability.

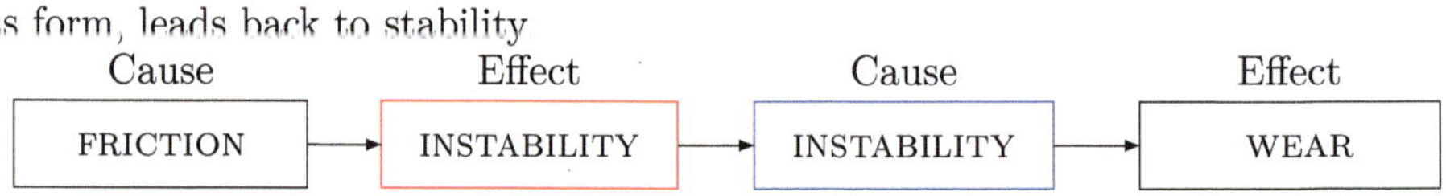

Figure 9: The altered cause-effect-chain

4 On the Phenomenon of Self-Organization

4.1 General Remarks

Self-organization is a term that is used in various ways across different sciences. The inconsistent usage has led to the emergence of different *theories of self-organization.* Examples of such theories include cybernetics, non-linear thermodynamics far from equilibrium, or the theory of autopoiesis. There are further examples as well.

Drawing on the work of I. PRIGOGINE [2], and also with an eye on the later focus of this book, what seems to unite these different *considerations of self-organization* is the existence of unstable states. [3] asserts, (self-organization) "is based ... on a nomological core: instabilities are constitutive."

For I. PRIGOGINE, the occurrence of instabilities is a necessary condition for triggering a self-organized process [2]. Lastly, a conceptual derivation shows that a system can organize itself and form structures without an external designer. Of course, the impetus is external, a triggering moment of instability. Very succinctly formulated by J.C. SCHMIDT [3]:"Wherever *self-organization* is discussed, phenomena of order formation and structure building are addressed." And later: "... how structures and order form *within* the system, without external specifications and external organizers."

In his works, I. PRIGOGINE elaborates on the phenomenon of *self-organization* and its relation to the theory of dissipative structures [24]. He recognizes that besides equilibrium, non-equilibria exist which can be the origin of order.

The formation of dissipative structures is linked to the presence of a nonequilibrium state. This is a fundamental condition "which is essential for understanding the coherence and organization in the world of nonequilibrium that surrounds us"[2]. Open dissipative systems are in exchange with their environment. An open system state allows for the export and import of entropy.

If the process conditions cause matter and energy flows, and thus structural changes, below a critical distance (i.e., close to equilibrium), so that the system does not reach the equilibrium state, it tends to approach an equilibrium-like state [25]. In this stationary nonequilibrium, external conditions prevent the system from reaching thermodynamic equilibrium. A state of *minimal entropy production* develops, which is close to equilibrium [24]. This principle of *minimal entropy production*, formulated by I. PRIGOGINE in 1945, applies to the maintenance of stationary states near equilibrium [25].

In [24], it is emphasized that not all changes are equivalent for a thermodynamic system.

$$dS = dS_e + S_i \tag{1}$$

The exchange term dS_e is of a completely different nature than the production term dS_i. The exchange of the system with the environment can occur in both directions. Thus, there is no definitive sign. In contrast, the internal entropy production can be 0 or > 0.

This difference in nature between the two terms on the right-hand side of Equation (1) is described by [2] as a difference in development towards an attractor state (equilibrium as an attractor) as opposed to a development initiated by external conditions.

At this point, I'd like to give additional information on the tribological process. Traditionally, in a friction pair, the occurring wear of the solid bodies is considered negative and disruptive. It is now evident that the system approaches a state due to irreversible entropy increase dS_i/dt, which it is somehow attracted to.

In the vicinity of equilibrium, a stationary situation can occur. If we further observe a continuous change of a parameter (e.g., a continuous sliding process), the system is pushed further away from equilibrium. It may reach a point where the system becomes *unstable* with respect to fluctuations. At that point, the state reached after this *bifurcation point* depends on the history of the system [2].

The occurrence of branching characterizes the principle that [2] calls *order through fluctuation*. An alternative emerges from a unique solution. The system has a choice [24], as shown in Figure (10). Further branches are possible thereafter.

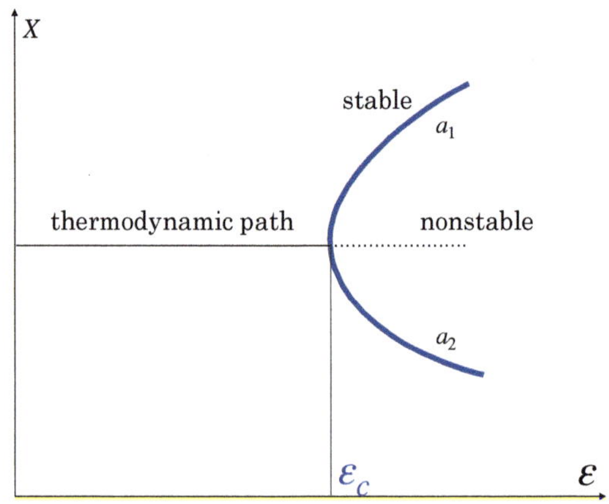

Figure 10: Diagram of a symmetric bifurcation after [2]. With ε as the control parameter, ε_c as the control parameter at the bifurcation point, X represents the thermodynamic path.

These fluctuations must exceed a critical magnitude far from equilibrium and can no longer be damped.

This critical magnitude is influenced by the system's ability to dampen fluctuations and the mechanisms that amplify these fluctuations within a particular domain [24]. To understand this critical magnitude, PRIGOGINE and STENGERS [24] write that this critical magnitude is based on the fact that the external world always seeks to dampen a fluctuation. Responsible for the destruction and amplification of an occurring fluctuation is the effectiveness of communication between the fluctuation region and the external world. "The critical magnitude measures the ratio between the volume in which the reaction occurs and the contact area through which this region interacts with the external world."

An almost spectacular example of the process of *self-organization* are the Bénard cells observed and named after H. BÉNARD. In a layer of liquid, a vertical temperature gradient and thus a continuous heat flow are generated. When this gradient reaches a critical value, the previously stationary state becomes unstable. The present mechanism of heat conduction transitions to convection, a collective movement of molecules that accelerates heat transport. An expression of this coherent movement is a new macroscopic structure consisting of hexagonal convection cells. Further examples are described in [24] and [3].

Figure 11: Heating a quantity of water filled with small rod-shaped elements. After a certain time, the elements arrange themselves into a honeycomb structure.

Taking a glance at the epistemological development over the centuries, the suspicion of system-immanent processes emerges early. SCHMIDT [3] also provides a comprehensive historical overview, which E.I. NKOYO picks up on in [25].

I would like to highlight the German philosophers I. KANT (1724 - 1804) and F.W.J. SCHELLING (1775 - 1854), who, albeit not conceptually, certainly had an idea of system-immanent process occurrences.[2]

In 1755, I. KANT [26] describes ideas about the formation of stars and planets (see Fig.(12)). From a filled space with smallest particles of matter, larger "clumps of matter" ultimately form through attraction and repulsion, up to the planets. He formulates sentences like "The elements ... are themselves a source of life". Or even more strikingly "... matter ... in its simplest state has a tendency to form a more perfect constitution

[2]The following paragraphs are reproduced with the original quotations in German at the end of the book.

through natural development". In describing the orbits of the formed bodies, one finds: "In this state..., since all particles ... move around the central body due to the acquired momentum, the conflict and the convergence of the elements is lifted, and everything is in the state of the smallest interaction."

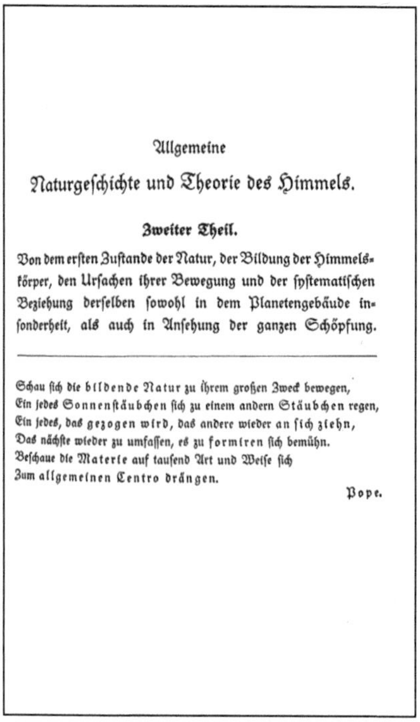
Allgemeine
Naturgeschichte und Theorie des Himmels.

Zweiter Theil.

Von dem ersten Zustande der Natur, der Bildung der Himmelskörper, den Ursachen ihrer Bewegung und der systematischen Beziehung derselben sowohl in dem Planetengebäude insonderheit, als auch in Ansehung der ganzen Schöpfung.

Schau sich die bildende Natur zu ihrem großen Zweck bewegen,
Ein jedes Sonnenstäubchen sich zu einem andern Stäubchen regen,
Ein jedes, das gezogen wird, das andere wieder an sich ziehn,
Das nächste wieder zu umfassen, es zu formiren sich bemühn.
Beschaue die Materie auf tausend Art und Weise sich
Zum allgemeinen Centro drängen.

Pope.

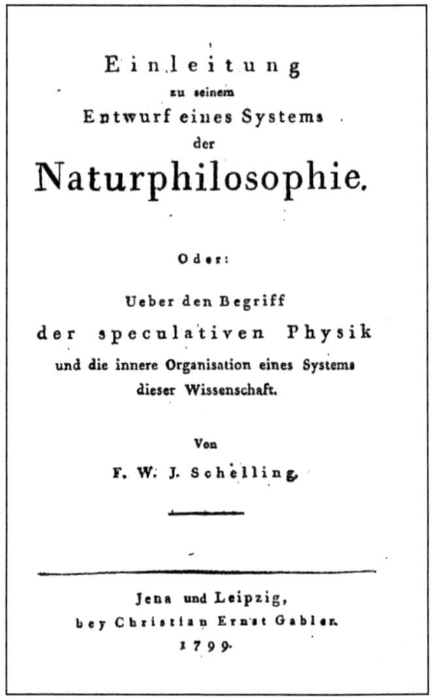
Einleitung
zu seinem
Entwurf eines Systems
der
Naturphilosophie.

Oder:

Ueber den Begriff
der speculativen Physik
und die innere Organisation eines Systems
dieser Wissenschaft.

Von
F. W. J. Schelling.

Jena und Leipzig,
bey Christian Ernst Gabler.
1799.

Figure 12: *left:* Title page of Kant's writing (1755) in an edition from 1925 and *right:* Title page of Schelling's writing from 1799

A generation later, SCHELLING writes in his Draft of a Philosophy of Nature from 1799 [27] (see Fig.(12)) in § VI,B,"... the secret of the production of *Nature from itself* lies in the sequence (of magnetism, electricity, and the chemical process), as it can also be distinguished in the individual body."And further in §3, B3, "... for this (Note: referring to nature) is Being or productivity itself."And earlier ".. Nature as a whole, both cause and effect of itself ..."can be found. Under § IV, A, "... that hovering of nature between productivity and product is thus perceived as a general duplicity of the principles by which nature must appear to be *in constant activity* (not italicized in the original) ...". He writes in §III "... that motion not only arises from motion, but also from rest, so that there is motion even in the stillness of nature ..". And finally, § II should be mentioned with "... Because all thought ultimately comes back to producing and reproducing, there is nothing impossible in the thought that the same activity by which nature reproduces itself anew in every moment ...".

Last but not least, in the 18th and early 19th centuries, we also find considerations by J.W. von Goethe that can be understood in relation to our present understanding of self-organization. He writes, ".. and one may therefore rightly assume an unstoppable progressive transformation"(cited in [28]).

4.2 The Concept of Entropy and Self-Organization in Postmodern Tribology

4.2.1 Introduction

Tribology presents itself as a discipline of the exact natural sciences [1] . In 1973, A. Knappwost further wrote :"Since the establishment of the two main laws of thermodynamics in the middle of the last century, it should have been recognized that ... the study of friction cannot be exhaustively treated solely on the basis of mechanics."And further:"All friction processes, however, form a special class of the so-called natural or irreversible processes...""Their common feature is therefore the *increase in entropy* of the entire system that occurs during their course."

At the beginning of this millennium, E.A. Assenova formulated [29]: "Tribology as the science of contact is rooted in the deep essence of things. The ability of a system to forget external disturbances is due to dissipation and under certain conditions leads to a specific organization of the system."

As early as the 1960s and 1970s, there were tribological studies whose interpretation led to the description of non-equilibrium states.

Early works by D.N. Garkunov et al. [30] concerning the so-called selective transfer, investigate tribological contacts as open thermodynamic systems. B.I. Kostjetzki et al. [31], [32] also describe structural-forming processes in their studies. Later, Kostjetzki explains [5]: "The diversity and variety of external friction processes create conditions for the emergence of a broad spectrum of *self-organization* states...""The general law consists of the following: For all materials and working media, there are ranges of loads and friction speeds within which the friction and wear parameters are stable and several orders of magnitude lower than outside these ranges."

In connection with fundamental studies on the description and definition of a tribological system, H. Czichos and Mølgaard [21], [22] mention the necessity of considering entropy.

In an assessment by A.A. Poljakov [33], it is stated: "The transition to open thermodynamic systems in friction, which are not destroyed, is a new stage in tribology and was made possible only through the evaluation of all experiences and the development of thermodynamics under non-equilibrium conditions."

It was B.E. Klamecki [34], [35], [36], who in the 1980s carried out an extensive analytical analysis of entropy production. He stated, among other things, that if the rate of energy input due to disturbances or continuous supply is greater than the rate of energy dissipation, the process loses its equilibrium state.

The process of self-organization is not a general, but a specific property of matter, define G. POLZER ET AL. [37].

In tribological studies by GERSHMAN ET AL. [38], it is noted that the process of *self-organization* is characterized by reduced entropy production, which in turn leads to a decrease in wear intensity. This is also a reason for the interest of tribologists in this concept.

M. NOSONOVSKY [39] states that wear is an irreversible change in the surface and thus leads to an increase in entropy. He suggests that it is reasonable to characterize wear with the concept of entropy.

Broadening this view, M.D. BRYANT [40] finds that wear is always associated with an irreversible permanent reorganization of a material's structure. He describes that any material transformation, including wear, is accompanied by the production of entropy.

In 2011, H. ABDEL-AAL [41] stated that "system failures, component damage, and deformations are to be regarded as byproducts of an adaptation process through which the system attempts to establish a stable equilibrium between the energy input, the effects of this input on the components of the subsystem, and the (immediate or remote) environment."

For a more comprehensive discussion beyond this introductory presentation, please refer to M. AMIRI and M.M. KHONSARI [42], M. NOSONOVSKY [39], and G.S. FOX-RABINOVICH and G.E. TOTTEN [43].

Studies of tribological processes involve considerations of natural driving forces and their effects within the so-called tribological system. I view friction as a natural process, the investigation of which leads us to fundamental questions about the phenomena surrounding us. Although we may not yet hope to answer these questions, the formulation of these questions represents a significant advance.

4.2.2 Selected Works Focusing on Solid-State Wear

H. ABDEL-AAL

In [41], H. ABDEL-AAL develops a concept of the open thermodynamic system for solid-solid contact. One side of the contact is illustrated in Figure (13a).

He balances the entropy production with an intrinsic term that considers, for example, plastic deformation, a term describing the heat conduction processes from the contact point into the material, and a term encompassing the material transport into and out of the system [44]. For an approximate application, three heat quantities are determined. These are the heat dissipated by friction in the contact, the heat conducted in the system, and the heat transport through material exchange. With $\dot{\sigma}$ representing the entropy

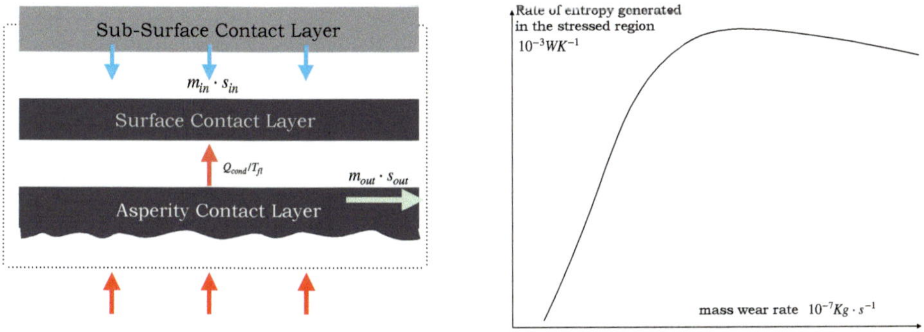

(a) Model of entropy transport 1

(b) Entropy rate und wear rate

Figure 13: Thermodynamic model and experimental wear investigation according to [41].

production, it follows [41]:

$$\dot{\sigma} = \frac{\lambda \dot{Q}_{gen}}{T} - \frac{\dot{Q}_{cond}}{T} + \sum_{in} \dot{m}s - \sum_{out} \dot{m}s \tag{2}$$

In this context, λ represents the proportion of heat flowing into a contact in a friction body. A parameter (ration residual entropy) is created that describes the material's ability to transport entropy and establishes a connection to the wear rate (Figure (13b)).

$$RRE = \frac{\lambda \dot{Q}_{gen} - \dot{Q}_{cond}}{\lambda \dot{Q}_{gen}} \tag{3}$$

M. Nosonovsky and B. Bushan

A consideration of entropy production on parallel and different scales is carried out by M. Nosonovsky and B. Bushan [45]. The idea pursued is that entropy is consumed on a macro-scale and produced on a micro-scale. They describe both a degradation and a healing process.

Initially, a concept of entropy production rate in the near-surface layer of a solid friction body is modeled. This is illustrated in Figure (14)[45]:

The entropy production rate is given by

$$\frac{dS}{dt} = \frac{(\mu W V)^2}{\lambda T^2} \tag{4}$$

Here, the original notation from [45] is retained. In this equation, μ represents the friction coefficient, W the normal force, V the relative velocity, λ the thermal conductivity, and T the temperature. Here's the translation:

The linear relationship between thermodynamic forces and flows

$$J_i = \sum_i L_{ki} Y_i \tag{5}$$

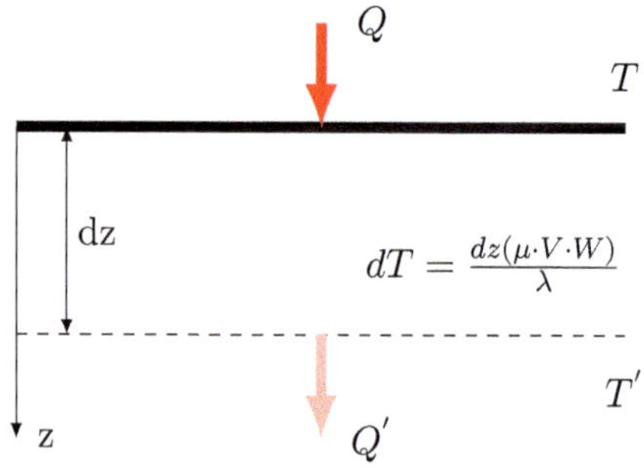

Figure 14: Heat flow in the near-surface region of a friction body

is solved as

$$J_{deg} = LY_{deg} + MY_{heal} \tag{6}$$

$$J_{heal} = MY_{deg} + HY_{heal} \tag{7}$$

where L, M, and H are the Onsager coefficients. This yields for the entropy rate

$$\dot{S} = \dot{S}_{deg} + \dot{S}_{heal} = \frac{L}{T}(Y_{deg})^2 + \frac{2M}{T} Y_{deg} Y_{heal} + \frac{H}{T}(Y_{heal})^2 \tag{8}$$

An example the authors investigate in [45] is the "healing"of cracks using filled microcapsules in the material. They write: "Some of this excess entropy can be consumed for the healing of bonds at the crack. The net entropy increases, however, not through the crack, but through the breaking of the microcapsules and an irreversible decrease in their number. Breaking the microcapsules puts the system into a non-equilibrium state, generating the restorative thermodynamic force Y_{heal}, which controls the flow and diffusion of the healing agent." Further examples are discussed in [45].

I.S. Gershman et al.

An interesting application of the concept of entropy and inherent system response (self-organization) is investigated by I.S. Gershman et al. in [46]. The application examines the contact between electrical lines during railway transportation. To assess the possibility of occurring instability, they utilize

$$\frac{1}{2} \frac{\partial}{\partial t}(\delta^2 s) = \sum_n \delta X_n \delta J_n \leqslant 0 \tag{9}$$

as a stability criterion according to [2], with $\delta X_h \delta J_h$ representing the so-called excess entropy. For just one independent friction process, they formulate the entropy production as

$$\frac{dS_i}{dt} = X_h J_h = \frac{(kpv)^2}{\lambda T^2} \tag{10}$$

Here, J_h denotes the heat flow, X_h represents the thermodynamic force generating the heat flow, corresponding to $gradT/T^2$, where $J_h = k \cdot p \cdot v$ with k being the friction coefficient, p the contact pressure, and v the relative velocity, and λ is the thermal conductivity.

Furthermore,

$$\frac{1}{2}\frac{\partial}{\partial t}(\delta^2 s) = \delta(kpv) \cdot \delta\left(\frac{kpv}{\lambda T^2}\right) = \frac{(pv)^2}{T^2}\left(\frac{1}{\lambda}\left(\frac{\partial k}{\partial \varphi}\right)^2 - \frac{k}{\lambda^2}\frac{\partial k}{\partial \varphi}\frac{\partial \lambda}{\partial \varphi}\right)\delta\varphi^2 \tag{11}$$

where a variable φ, describing the distance from equilibrium, has been introduced, yielding the interpretation

$$\frac{\partial k}{\partial \phi}\frac{\partial \lambda}{\partial \phi} > 0 \tag{12}$$

for there to be a probability of instability, allowing the right side of equation (11) to become negative. In [46], the occurrence of condition (12) is explained with examples (also see [38], [47]).

M.D. Bryant, M.M. Khonsari, and F.F. Ling

In a fundamental paper, M.D. Bryant, M.M. Khonsari, and F.F. Ling [48] describe the essential correlation between degradation processes in a system and entropy production. They develop the so-called *DEG - degradation entropy production theorem.* It is natural to relate this universally applicable theorem to irreversible tribological wear [49], [50]. In doing so, a degradation coefficient $\boldsymbol{B}$ is formulated, correlating wear (as an irreversible degradation process) with the corresponding entropy production.

In general, Bryant et al. describe a degradation process that consists of $i = 1, 2, ...n$ dissipative processes p_i. Each process $p_i = p_i(\zeta_i^j)$ describes an energy, work, or heat property and depends on a series of time-dependent phenomenological variables $\zeta_i^j = \zeta_i^j(t), j = 1, 2, ...m$. A degradation quantity w is defined as

$$w = w\left[p_i\left(\zeta_i^j\right)\right], \quad i = 1, 2, ...n; \quad j = 1, 2, ...m_i \tag{13}$$

which depends on the phenomenological variable ζ_i^j across the n processes p_i. It is clear that each dissipative process produces entropy $S_i' = S_i'\left[p_i\left(\zeta_i^j\right)\right]$ (original notation from [48]) and is characterized by the same variables ζ_i^j. Now, a degradation rate $\dot{w}$ is formed as

$$\dot{w} = \frac{dw}{dt} = \sum_i\sum_j\left(\frac{\partial w}{\partial p_i}\frac{\partial p_i}{\partial \zeta_i^j}\right)\frac{\partial \zeta_i^j}{\partial t} = \sum_i \dot{w}_i = \sum_i\sum_j Y_i^j J_i^j \tag{14}$$

and for the entropy production rate

$$\dot{S}' = \frac{dS'}{dt} = \sum_i\sum_j\left(\frac{\partial S'}{\partial p_i}\frac{\partial p_i}{\partial \zeta_i^j}\right)\frac{\partial \zeta_i^j}{\partial t} = \sum_i \dot{S}_i' = \sum_i\sum_j X_i^j J_i^j \tag{15}$$

For stationary systems or systems close to equilibrium, the entropy production is given by

$$S_i' = \sum_j X_i^j \cdot J_i^j \tag{16}$$

As the product of a thermodynamic force and a thermodynamic flux. Comparing equations (14) and (15), the authors of [48] analogously describe a general degradation force Y_i^j akin to a general thermodynamic force X_i^j, and from this, they form the degradation coefficient $\boldsymbol{B}$ as

$$\boldsymbol{B}_i = \frac{Y_i^j}{X_i^j} = \frac{(\partial w/\partial p_i)(\partial p_i/\partial \zeta_i^j)}{(\partial S'/\partial p_i)(\partial p_i/\partial \zeta_i^j)} = \frac{\partial w/\partial p_i}{\partial S'/\partial p_i} = \frac{\partial w}{\partial S'}|_{p_i} \tag{17}$$

For an active process p_i, equation (17) illustrates the interaction between entropy production and degradation. An application to wear [49] demonstrates the degradation coefficient $\boldsymbol{B}$ as the slope in a diagram of degradation (here, solid-state wear) versus entropy production. Extended and in-depth discussions can be found in [40], [48], [49], [50], [51] as well as in [52], [53], [54], [55].

A.B. Amiri and M.M. Khonsari

A.B. Amiri and M.M. Khonsari[55] propose two coefficients for the consideration of solid-state wear. They introduce the so-called WED coefficient, which relates power loss and wear, and the PTR coefficient as the power loss temperature rise coefficient. They demonstrate through experimental examples the practicality of working with these coefficients.

It holds that

$$\dot{w} = \psi_w \cdot P_d \tag{18}$$

$$P_d = \psi_T \cdot \triangle T \tag{19}$$

where $\dot{w}$ is the wear rate, ψ_w is the WED coefficient, P_d is the dissipated frictional energy, ψ_T is the PTR coefficient, and $\triangle T$ is the temperature rise. The connection to the DEG theorem provides

$$\boldsymbol{B} = \psi_w \cdot T \tag{20}$$

A comparative study between the wear factor k and the degradation coefficient $\boldsymbol{B}$ is conducted by K.P. Lijesh and M.M. Khonsari in [56] , demonstrating the advantages of using $\boldsymbol{B}$.

Y.P. Kozyrev and E.B. Sedakova

In the case where a self-organization process is triggered, Y.P. Kozyrev and E.B. Sedakova [57] investigate the wear behavior of a selected friction pair. They consider a friction term (heat conduction) and a diffusion term for describing entropy production as follows:

$$\frac{dS_i}{dt} = X_1 J_1 + X_2 J_2 \tag{21}$$

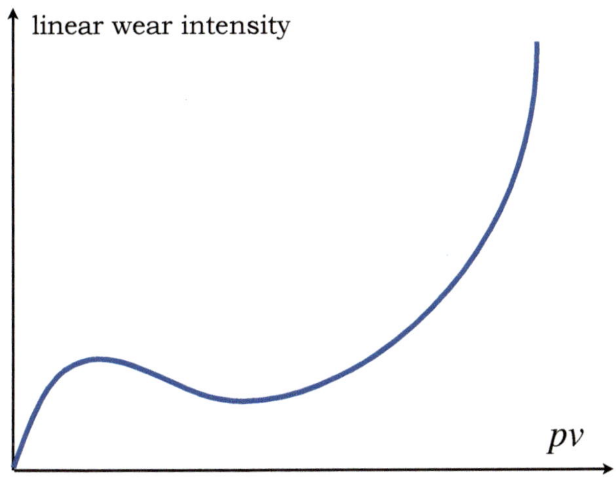

Figure 15: Wear investigation with self-organization behavior according to [57]

$$\frac{dS_i}{dt} = \frac{(fpv)^2}{\lambda T^2} + \frac{\gamma_D(\mathrm{grad}\varphi)^2}{T} \tag{22}$$

(in the original notation), where $J_1 = -\lambda \mathrm{grad} T = fpv$, with p being the contact pressure, v the relative velocity, f the friction coefficient, λ the thermal conductivity, φ the chemical potential, and γ_D the transport coefficient. Furthermore, $J_2 = -\gamma_D \mathrm{grad}\varphi$ and $X_2 = \mathrm{grad}\varphi/T$. They relate the transport coefficient to wear and obtain

$$\gamma_D = \gamma_{D0} - \frac{f(pv)^2}{\lambda T(\mathrm{grad}\varphi)^2} \tag{23}$$

It is evident that with an increase in stress (pv), the parameter γ_D (here indicative of wear) decreases (indicating a steady state). Experiments confirm the behavior described by equation (23) (as shown in Figure (15)).

5 Postmodern Lubricating Grease Tribology

5.1 Introduction (Entropy Concept, Self-organization)

Greases are colloidal-disperse systems with distinctly viscoelastic properties. At the heart of this chapter and also this book is the *frictionally stressed grease*, in stark contrast to a *grease*-lubricated friction pair.

In the conception, a tribological subsystem is developed into which mechanical energy is introduced. Observing the effect of this disturbance demonstrates in various ways the *irreversible* nature of the processes occurring, the system's quest for ways of energy dissipation, and the endeavor of all real processes to reach a stable state.

Quite evidently, even in this particular system, the central role of emerging *instability* is apparent. *Instability* serves as the source and starting point for highly differentiated behavior. Proximity to and distance from equilibrium are important criteria for the system's response, and typically, the stressed grease volume element is an open system. That is, we observe energy and matter exchange.

The perspective on wear formulated in this book also leads to a cause-and-effect chain for the *stressed grease* system that is expanded compared to traditional tribology.

If continuous energy input makes a return to equilibrium, and thus stability, impossible, a *stationary non-equilibrium* becomes an attractive stable process state with minimal entropy production.

Stability⟶Friction⟶**Instability**⟶Wear⟶**Stability**

Figure 16: The altered cause-effect-chain as also shown in Fig.(9)

The altered cause-effect-chain:

Friction is not the direct cause of wear, but rather the cause of an instability-triggering disturbance through the input of energy into the system [58].

The state far from equilibrium harbors the possibility of forming new structures - a special phenomenon that the system *itself* initiates. PRIGOGINE [2] calls the self-organized structures *dissipative structures.* It is suspected, and indirect observations in experimental work suggest, that the process of *self-organization* can also occur in stressed lubricating grease films. Direct observation is currently hardly conceivable and was not possible within the framework of the investigations described here.

The introduction of the entropy concept into lubricating grease tribology in 2010 [59] opened up the possibility of accounting for the irreversible nature of the process, considering unstable states, and understanding the driving forces of the process a little better.

The modeling of the stressed lubricating grease volume element as an open thermodynamic system is shown in Figure (17).

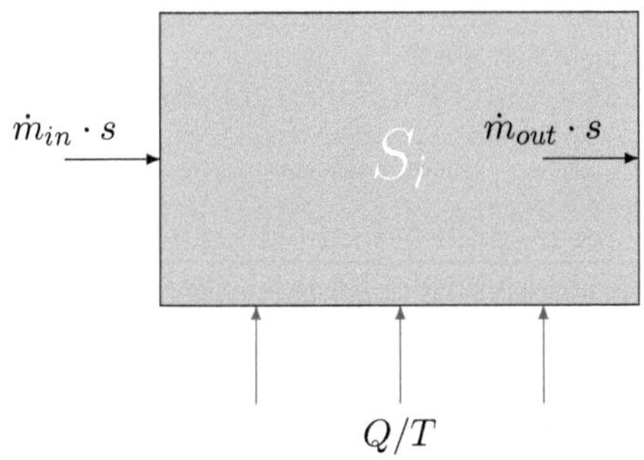

Figure 17: Entropy transport in the open thermodynamic system of stressed lubricating grease

The term $(\dot{m}_{in} \cdot s)$ represents the entropy entering the system with the unstressed structure, $(\dot{m}_{out} \cdot s)$ the entropy leaving the system with the changed (weared) structure, Q/T the entropy transported into the system via heat, m the mass of the transported structure, s the specific entropy, and S_i the entropy produced within the system. Simplifying, the figure shows a volume element that experiences a friction process on only one side.

5.2 Investigations in Stationary Non-Equilibrium

5.2.1 On Entropy Transport

The focus is initially on the energetic stressability of the modeled system, observing the irreversible process through the system entropy. In these investigations, the input of mechanical energy is considered, while other processes such as oxidation, chemical activation, etc., are not taken into account. Figure (17) also shows the possibilities for changing the system entropy. The following applies:

$$dS = dS_i + dS_e \tag{24}$$

with the two terms S_i - the entropy production, and S_e - the entropy transport. Entropy production occurs within the defined thermodynamic system. The change due to entropy transport occurs across the system boundaries as an exchange of energy and matter with the surroundings.

The following applies:

$$dS_i \geqslant 0 \quad \text{and} \quad dS_e \lesseqgtr 0 \tag{25}$$

The state of thermodynamic equilibrium is characterized by $S = \max$ and $dS = 0$. Entropy production becomes zero when no process mechanisms occur within the system.

In local form, Equation (24) can be written as [60]:

$$\frac{\partial \rho s}{\partial t} = -\text{div} J_s + \sigma \tag{26}$$

$$\sigma \geqslant 0 \tag{27}$$

Here, ρs is the entropy density, and σ is a source term (production) that satisfies the inequality (27) (see also [61]). The prerequisite is that "the local macroscopic measurements performed on a system are indeed measurements of the properties of a small part of the system, a part that still contains a sufficiently large number of the components that make up the system"[60].

Figure (17) shows a system with both matter and heat flow, and thus with inflowing and outflowing entropy, alongside entropy production within the system boundaries. For multiple matter streams, the following equation applies [62]:

$$\frac{dS}{dt} = \dot{S}_Q + \dot{S}_i + \sum_{in} s_{in}\dot{m}_{in} - \sum_{out} s_{out}\dot{m}_{out} \tag{28}$$

First, the influence of entropy transport on the energetic usability will be examined. For this purpose, the ratio of the applied friction energy to the volume of the altered solid structure (thickener structure) will be used. This ratio corresponds to an energy density and is called the apparent friction energy density, analogous to [63], and is considered in the *stationary non-equilibrium.* This also means that all occurring quantities are constant over time. From Equation (28), the following can now be written [64]:

$$\boxed{e^*_{Rrheo} = T_f \cdot (\rho_{out} \cdot s_{out}) - \frac{T_f}{V_{out}}(\dot{m}_{in} \cdot s_{in} + \dot{S}_Q)} \tag{29}$$

The term $(\rho_{out} \cdot s_{out})$ can be interpreted as an entropy density that exits the system with the altered structure. It is interesting to note that there is a direct proportionality between the level of energetic strain on the system and the entropy density exiting the system, which is coupled to the altered structure. The following conclusion can be drawn from this:

A comparatively high entropy density, which is coupled to the matter exiting the system, enables this system to be energetically strained again. To eliminate the disturbance (in this case, friction energy), a small volume of altered structure (lubricating grease) is sufficient when the entropy density is high. In contrast, a system that can only couple a low entropy density requires a larger structural volume. Similarly, this concept applies to the abraded solid wear volume.

Such different system behavior were observed in rheometer experiments with selected grease samples. Lubricating greases were examined using a three-stage test procedure on the rheometer. A sample with Li soap (NLGI2) and a sample with a polyurea thickener (NLGI2) were first sheared and then immediately examined with an amplitude sweep [65].

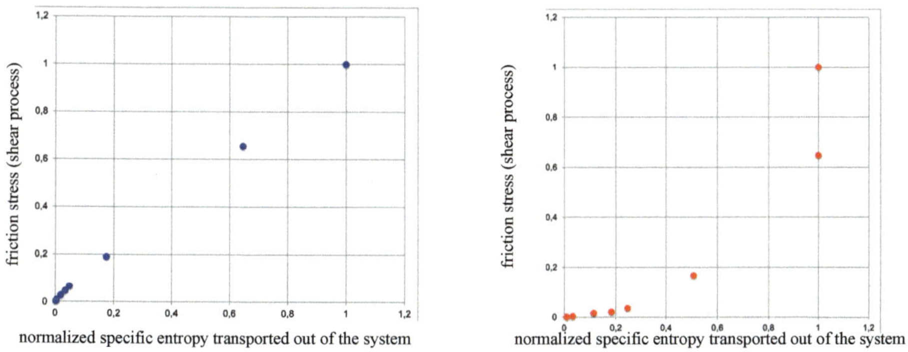

Figure 18: Entropy transport *left*: Li sample and *right*: PU sample

In Figure (18), the ordinate represents the frictional stress, i.e., shear at different shear rates, and the abscissa shows the specific entropy that leaves the system.

Both systems behave very differently in the range of lower stresses. Comparatively, the PU sample (right) responds with a stronger transport of specific entropy through the altered structure than the Li sample.

When examining the wear behavior, i.e., the intensity of structural changes, the behavior is illustrated in Figure (19).

In Figure (19), the ordinate shows the energy required in the amplitude sweep after the shear stress to reach the crossover point. High values indicate a structure that is little changed by the previous shear, while lower values suggest relatively high wear of the lubricating grease. Of course, this is a comparative analysis based on a defined test procedure.

This representation also clearly shows the different behaviors of the two samples. In the range of lower frictional stresses (shearing at lower $\dot{\gamma}$), the Li sample (•) responds with intensive structural degradation. In comparison, the PU sample (•) exhibits very minimal lubricating grease wear. Considering all three illustrations, we can see different reactions of the system to energetic stress. The Li sample responds with increased structural changes and comparatively lower entropy transport, while the PU sample achieves high entropy transport and thus requires less lubricating grease wear. These are two different methods that both serve to relieve energetic stress. Both samples were of NLGI class 2, consisting exclusively of a PAO base oil and a thickener.

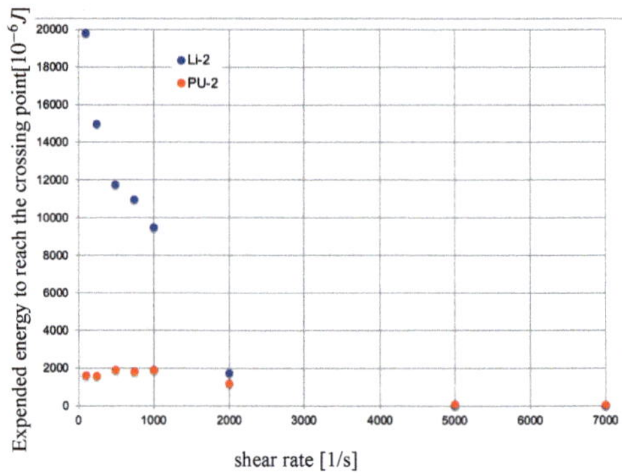

Figure 19: Wear behavior of lubricating grease under different stresses after test procedure

5.2.2 Process and Counterprocess in Stationary Non-Equilibrium

When a lubricating grease sample is subjected to a constant shear process and the resulting shear stress is observed, one obtains the well-known flow curve τ vs. t with $\dot{\gamma} = \text{constant}$. This curve shows a time-dependent initial part that transitions into an approximately constant behavior (Figure (20)). The system then reaches a state of *stationary non-equilibrium.*

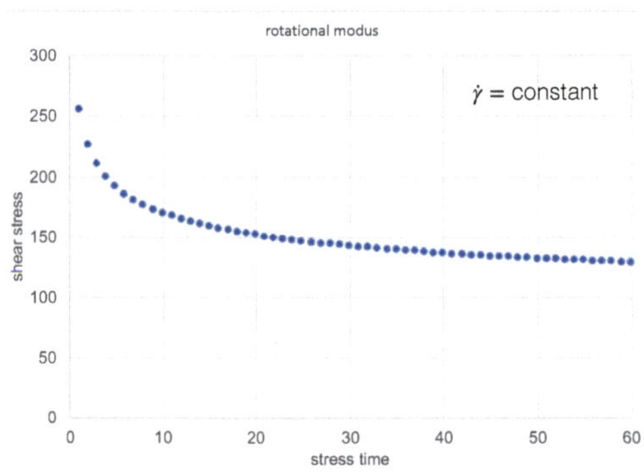

Figure 20: Shear stress behavior at constant shear rate and constant temperature

In this state, processes continue to occur with minimal entropy production. One can imagine an ongoing fragmentation of macroscopic agglomerates and coagulation through the collision of micro-geometric particles and fibrils. These mechanisms are also indicated by R. Czarny [66] and J.M. Franco (personal communication). Two opposing processes are observed that balance each other out.

This system is subsequently described based on the work of M. Nosonovsky and B.

BUSHAN [45], who develop a concept where processes occur simultaneously at macro and micro levels. They refer to degradation and healing processes (see also the description in section (4)).

Fragmentation and coagulation are now described in a system of equations developed with simple assumptions [67]. The fragmentation process is associated with a degradation parameter ξ, while coagulation, caused by collision, is associated with the parameter ζ.

In non-equilibrium thermodynamics, thermodynamic forces X and thermodynamic fluxes J are related by:

$$J_k = \sum_i L_{ki} X_i \tag{30}$$

where L_{ki} is the Onsager coefficient. Based on the reciprocity relation, $L_{ki} = L_{ik}$ [60]. The forces X_{deg} and X_{heal} are considered as external forces, leading to:

$$J_{deg} = K \cdot X_{deg} + M \cdot X_{heal} \tag{31}$$

$$J_{heal} = M \cdot X_{deg} + H \cdot X_{heal} \tag{32}$$

Here, K, M, and H are the Onsager coefficients. The degradation process refers to fragmentation, and the healing process refers to coagulation. It is further assumed that the degradation process and the healing process increase when positive forces are applied, and both processes weaken when opposite forces are applied. This implies $K > 0$, $H > 0$, and $M < 0$.

In accordance with [39], simplifying conditions are postulated. It is assumed that there is a constant degradation force with $X_{deg} = \rho$. Furthermore, a healing force proportional to the healing parameter is assumed, $X_{heal} = f(\xi)\rho$. Here, it is assumed that the coagulation rate depends on the fragmentation process, with $f(\xi)\rho$, owing to the fact that some of the energy dissipates into the oil film during particle collision. A constant dependence is assumed, $f(\xi) = -\omega$. Thus, $X_{heal} = -\omega\zeta$, and an equal balance of fragmentation and coagulation is assumed. This leads to:

$$\dot{\xi} = -K\omega\zeta - M\omega\zeta \tag{33}$$

$$\dot{\zeta} = -M\omega\zeta - H\omega\zeta \tag{34}$$

The solution to this system of differential equations yields:

$$\xi = a \cdot e^{-\omega(M+H)t} + b \tag{35}$$

$$\zeta = a \cdot \frac{M+H}{K+M} \cdot e^{-\omega(M+H)t} \tag{36}$$

Here, K is associated with the fragmentation process and H with the coagulation process. In the case where $K = H$, the functions are identical but shifted by b. I interpret this as indicating that the structural change, or rather the formed structure, never falls below the level of the pure base oil.

5.2.3 Application of the DEG Theorem to Lubricating Grease Wear

For this investigation, let's consider the observed system as closed and in a steady state. Initially, a degradation process $w = w[p(\zeta)]$ describes a single process $p(\zeta)$ with a phenomenological variable ζ [65]. Here, the degradation process refers to lubricating grease wear, which manifests as a structural change. It is expressed as:

$$P_{St} = P_{St}(E_f(P_E)) \tag{37}$$

representing the dissipation of frictional energy E_f (see Definition (2.11)), caused by an energy input P_E. The entropy production process is also influenced by the time-dependent variable (P_E) and is denoted as:

$$S_i = S_i(E_f(P_E)) \tag{38}$$

For the entropy production rate:

$$\frac{dS_i}{dt} = \left(\frac{dS_i}{dE_f} \cdot \frac{dE_f}{dP_E}\right) \cdot \left(\frac{dP_E}{dt}\right) = X \cdot J \tag{39}$$

And for the rate dP_{St}/dt:

$$\frac{dP_{St}}{dS_i} = \left(\frac{dP_{St}}{dE_f} \cdot \frac{dE_f}{dP_E}\right) \cdot \left(\frac{dP_e}{dt}\right) = Y \cdot J \tag{40}$$

If we take the ratio of the lubricating grease wear rate dP_{St}/dt to the entropy production rate dS_i/dt, we have:

$$\frac{dP_{St}/dt}{dS_i/dt} = \frac{dP_{St} \cdot dE_f \cdot dP_E}{dE_f \cdot dP_E \cdot dt} \cdot \frac{dE_f \cdot dP_E \cdot dt}{dS_i \cdot dE_f \cdot dP_E} = \frac{dP_{St}/dt}{dS_i/dt} = \frac{Y}{X} = \boldsymbol{B} \tag{41}$$

Here, $\boldsymbol{B}$ represents the degradation coefficient according to BRYANT ET AL.. Therefore:

$$\boldsymbol{B} = \frac{Y}{X} = \frac{dP_{St}}{dS_i} \tag{42}$$

$\boldsymbol{B}$ indicates the correlation between structural degradation and entropy production. Similarly, we have:

$$\frac{dP_{St}}{dt} = \boldsymbol{B} \cdot X \cdot J = \boldsymbol{B} \cdot \frac{dS_i}{dt} \tag{43}$$

For this investigation, lubricating grease wear will be considered differently from previous approaches. Neither an indirect assessment via the shear stress curve nor an indirect assessment from an oscillation measurement will be used. Instead, an attempt will be made to incorporate the geometry (size) and the change in the number of solid particles (Figure (21)).

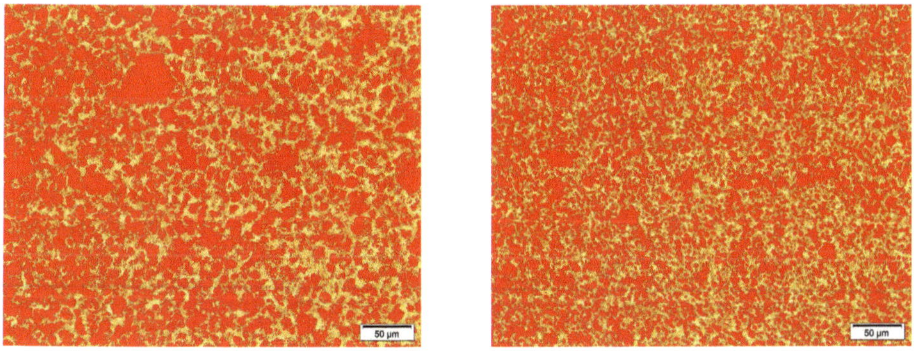

Figure 21: Transmitted light microscope image: *left* distribution of unstressed agglomerates, *right* distribution of stressed agglomerates in the rheometer

The illustration in Figure (22) is intended to depict the assumed change in particle count triggered by a frictional process.

It is postulated that:

$$\triangle E_f \sim \triangle n_p \tag{44}$$

meaning that the change in frictional energy $\triangle E_f$ leads to a change in particle count n_p. We express this as:

$$dE_f = E_p \cdot dn_p \tag{45}$$

Here, we introduce a proportionality factor E_p. Its nature becomes evident below with the DEG theorem. In terms of time:

$$\frac{dE_f}{dt} = E_p \cdot \frac{dn_p}{dt} \tag{46}$$

Thus, for the temporal change in particle count n_p:

$$\frac{dn_p}{dt} = \frac{\tau \cdot V}{E_p} \cdot \frac{d\gamma}{dt} \tag{47}$$

Equation (47) expresses the lubricating grease wear through the change in particle count n_p.

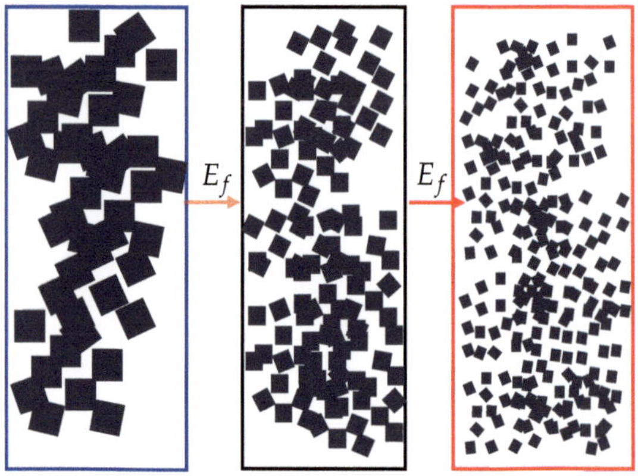

Figure 22: Illustrated representation of the change in particle count due to wear-inducing frictional process

With the starting equation (45) and the entropy production rate according to equation (39), assuming complete dissipation of the frictional energy:

$$\frac{dS_i}{dt} = \left(\frac{dS_i}{dE_f} \cdot \frac{dE_f}{dP_E}\right) \cdot \left(\frac{dP_E}{dt}\right) \tag{48}$$

Further, following [48], for a closed system in a steady state, the transformation $dS_i/dE_f = 1/T$ can be applied. Let's set the time-dependent phenomenological variable $\zeta = P_E = \gamma$ (deformation) and write:

$$\frac{dE_f}{d\gamma} = \tau \cdot V \tag{49}$$

It follows that:

$$\frac{dP_{St}}{dt} = Y \cdot J = \boldsymbol{B} \cdot X \cdot J \tag{50}$$

Thus:

$$\frac{dn_p}{dt} = \frac{\boldsymbol{B} \cdot \tau \cdot V}{T} \cdot \frac{d\gamma}{dt} \tag{51}$$

Considering equation (46), we have:

$$\frac{dn_p}{dt} = \frac{\tau \cdot V}{E_p} \cdot \frac{d\gamma}{dt} \tag{52}$$

Comparing equations (51) and (52), we can identify the proportionality factor as:

$$E_p = \frac{T}{\boldsymbol{B}} \tag{53}$$

Figure (23) provides an example with selected $\boldsymbol{B}$ values.

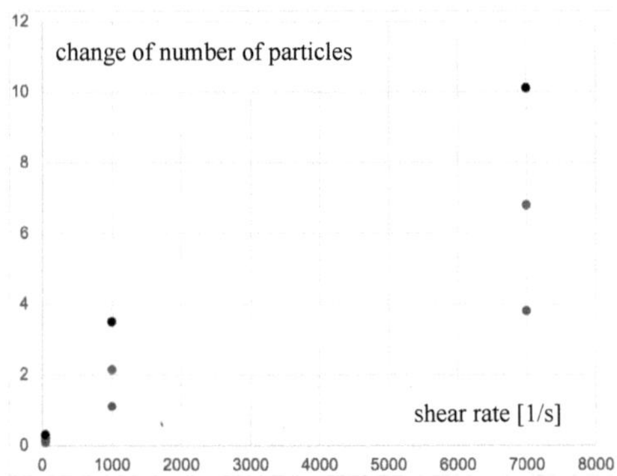

Figure 23: Change in lubricating grease wear with assumed $\boldsymbol{B}$ values: 1.5, 1, 0.5.

In [40], it is pointed out that the degradation coefficient is the slope of a curve representing the degradation versus entropy production.

5.3 On the Possibility of Self-Organization in Lubricating Grease Stress

5.3.1 Conditions for Possible Formation of New Structures

The starting point of the investigations in this section is the fact that after the occurrence of instabilities, a different, new structure can form. This new structure then leads the system to stability.

According to the definition (2.15), the lubricating grease structure includes, among other things, the distribution of solid particles in the base oil. It is suspected that a possibility of structure formation could lie in this distribution of solid particles. This means that a geometric or temporal structure may form with an increased dissipation rate, initiating stable processes.

To consider these possibilities, the conditions for self-organization need to be analyzed. The process is examined with regard to the occurrence of instabilities, which are a prerequisite for triggering self-organized events.

If wear is understood as the possibility to dissipate frictional energy (see also Definition (2.11)), i.e., to eliminate or reduce an energetic disturbance, the concepts of self-organization can also be applied to the system of *lubricating grease volume under stress.* Some examples are found in Table (5.1).

It has been extensively described that a *self-organization process* can be initiated when the system has lost its stability. Thus, when disturbances far from thermodynamic equilibrium are no longer damped, but instead lead to a change in the degree of order and ultimately enable a stable process flow.

Effect	Driving Force	Condition to Initiate	Final Configuration
Stationary micrography	Feedback due to coupling of friction and wear	Wear effects	Minimum of friction and wear
In situ tribofilm	Chemical reaction leads to film growth	Wear decreases with increasing film thickness	Minimum friction and wear at stationary film thickness
Slip waves	Dynamic instability	Unstable sliding	Reduction of friction
Self-lubrication	Embedded self-lubrication mechanism	Thermodynamic criteria	Reduced friction and wear
Surface-healing	Embedded self-healing mechanism	Proper coupling of degradation and healing	Reduced wear
Stationary lubricating grease structure	Involved fragmentation and coagulation	Thermodynamic criteria	Minimum entropy production

Table 5.1: Effects of self-organization according to [39] and extended from [67]

The initiation of self-organized structure formation is contingent upon conditions that give rise to an instability. This needs to be analyzed, and according to PRIGOGINE [2], the possibility of an unstable process behavior is investigated using the term $\delta^2 S$ as a LYAPUNOV function. Generally, this means examining the development of a system and addressing the question of whether the system moves toward the equilibrium point. The equilibrium point appears as an attractor if the time derivative of the overall process behavior has the opposite sign of the overall behavior. An illustration is also provided by [2] in Figure (24).

It appears evident that entropy production for non-equilibrium processes can be regarded as a LYAPUNOV function . Entropy production will increase in response to a disturbance in the system and will then return to the entropy production minimum as an internal reaction.

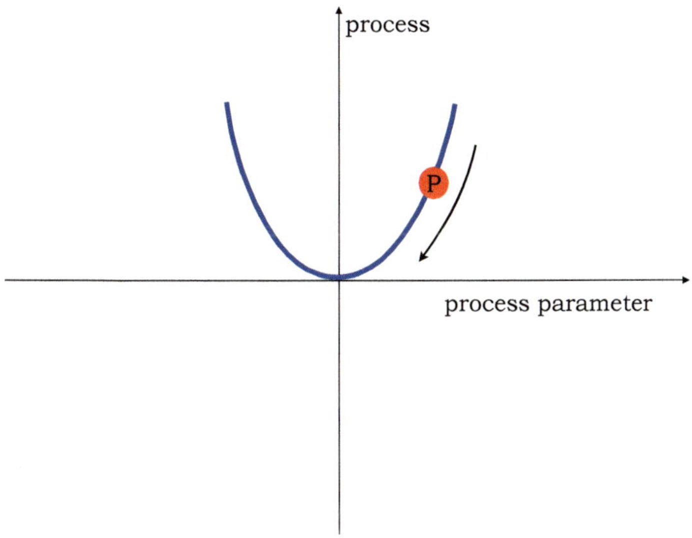

Figure 24: A disturbed system pushed to point P responds by developing toward the equilibrium point (according to PRIGOGINE [2])

When a system near equilibrium is disturbed, the entropy is given by

$$S = S_e + \delta S + \frac{1}{2}\delta^2 S \tag{54}$$

where S_e is the equilibrium entropy and δS is the first-order term, which vanishes at equilibrium or at maximum entropy. Therefore, the stability of the system is determined by the sign of the second-order term $\delta^2 S$ [2], [68], [69], [61].

In [69], one finds: "A stationary state of a system is stable if the associated negative-definite quantity $\delta^2 S$ is a monotonically increasing (i.e., monotonically approaching zero) function of time for all times ($t > t_0$) after the occurrence of the disturbance." The behavior is illustrated in Figure (25). It shows a negatively monotonically increasing function, which $= 0$ when the disturbance is removed. In [2], the connection between the time derivative

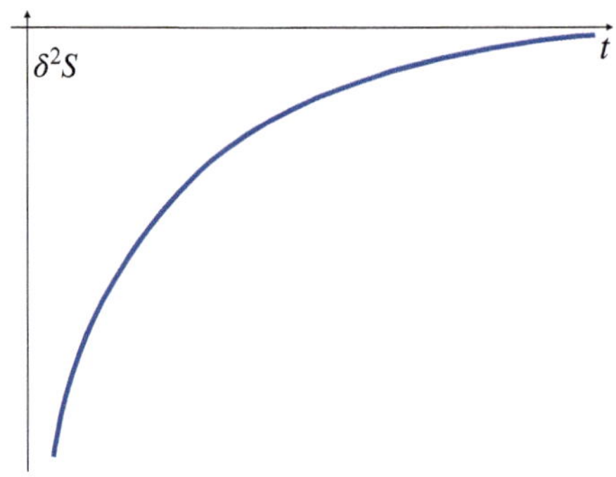

Figure 25: Behavior of the function $\delta^2 S$ according to [69]

of $\delta^2 S$ and entropy production is given by

$$\frac{1}{2}\frac{\partial}{\partial t}\delta^2 S = \sum_j J_j X_j = P \geqslant 0 \tag{55}$$

with

$$P \equiv \frac{d_i S}{dt} \geqslant 0 \tag{56}$$

To examine stability in regions far from equilibrium, the disturbance $\delta^2 S$ is studied near non-equilibrium. Here, instead of analyzing the forces X and fluxes J in the stationary region, their disturbances are investigated. The excess entropy production is observed. As GLANSDORFF and PRIGOGINE demonstrate [70], if

$$\frac{1}{2}\frac{\partial}{\partial t}\delta^2 S = \sum_j \delta J_j \delta X_j \geqslant 0 \tag{57}$$

holds for all $t > t_0$, $\delta^2 S$ is a LYAPUNOV function and stability is ensured. Here, t_0 is the time when the disturbance begins.

5.3.2 Studies on Self-Organization in Friction-Stressed Lubricating Grease

Since direct experimental observations are lacking, which could be interpreted as formed dissipative structures, indirect evidence is sought. These are indications that suggest the likely formation of new levels of order. Examples include transitions between high and low friction levels in friction experiments or contrasting wear behaviors, among others. The occurrence of instabilities can then be seen as signposts for possible self-organization processes.

Initially, the general case of friction in the grease film will be examined. As previously mentioned, the entropy production can be expressed as follows:

$$\frac{dS_i}{dt} = X_h \cdot J_h \tag{58}$$

Here, X_h is the thermodynamic force that initiates the heat flow, and J_h is the thermodynamic flux. According to GROOT and MAZUR [60], we have

$$X_h = \frac{\operatorname{grad} T}{T^2} \tag{59}$$

and

$$J_h = -\lambda \cdot \operatorname{grad} T \tag{60}$$

Introducing the total frictional energy per unit time and $J_h = \tau \cdot \dot{\gamma} \cdot V$ [71], the entropy production is then given by

$$\frac{dS_i}{dt} = X_h \cdot J_h = \frac{(\tau \cdot \dot{\gamma} \cdot V)^2}{-\lambda \cdot T^2} \tag{61}$$

Here, τ represents the shear stress, $\dot{\gamma}$ the shear rate, V the volume under stress, λ the thermal conductivity, and T the temperature. Now, let's examine the deviations from the thermodynamic forces and fluxes in the steady state.

$$\frac{1}{2}\frac{\partial}{\partial t}(\delta^2 S) = \delta(\tau \cdot \dot{\gamma} \cdot V)\delta\left(\frac{\tau \cdot \dot{\gamma} \cdot V}{-\lambda \cdot T^2}\right) \tag{62}$$

This implies the investigation of excess entropy production. Following a similar approach to [46], we introduce a parameter ε describing the distance from thermodynamic equilibrium. We consider the dependencies $\tau(\epsilon)$ and $\lambda(\epsilon)$. Consequently,

$$\frac{1}{2}\frac{\partial}{\partial t}(\delta^2 S) = \delta(\tau \cdot \dot{\gamma} \cdot V)\delta\left(\frac{\tau \cdot \dot{\gamma} \cdot V}{-\lambda \cdot T^2}\right) = \frac{\dot{\gamma}^2 \cdot V^2}{T}\left[\frac{1}{\lambda}\left(\left(\frac{\partial \tau}{\partial \varepsilon}\right)^2 - \frac{\tau}{\lambda^2}\frac{\partial \tau}{\partial \varepsilon}\frac{\partial \lambda}{\partial \varepsilon}\right)\delta\epsilon^2\right] \tag{63}$$

Now, we can search for conditions that would make the right hand side of equation (63) negative, indicating system instability. To enable instability, shear stress and thermal conductivity must increase or decrease as ϵ grows. This implies:

$$\frac{\partial \tau}{\partial \varepsilon}\frac{\partial \lambda}{\partial \varepsilon} > 0 \tag{64}$$

This seems feasible when considering a rheometer experiment. A constant shear rate, due to internal energy dissipation, leads to the well-known shear stress curve. At the same time, the decrease in agglomerate size, with their oil coatings, results in decreased thermal conductivity. Therefore, the formation of new dissipative structures is indeed conceivable.

In a further step, let's attempt a detailed observation. Three mechanisms are selected:

- General shearing of the base oil
- Fragmentation of solid particle agglomerates
- Coagulation of the agglomerates

Other mechanisms conceivable in the friction process are not initially investigated. For entropy production, we have:

$$\frac{dS_i}{dt} = \frac{dS_{i(\text{oil})}}{dt} + \frac{dS_{i(\text{frag})}}{dt} + \frac{dS_{i(\text{coag})}}{dt} \tag{65}$$

For the fragmentation process, the rate of entropy production influences the energy required to overcome a critical deformation. For the coagulation process, the kinetic energy of the flowing agglomerates plays a role.

$$\frac{dS_i}{dt} = \frac{(\tau_{(\text{oil})} \cdot \dot{\gamma} \cdot V)^2}{-\lambda \cdot T^2} + \frac{\gamma^2_{\text{critic}} \cdot G' \cdot \cos\delta^{-1} \cdot V \cdot F_f}{T} + \frac{[E_{\text{kin(b)}} - (E_{\text{kin(a)}} + E_{\text{def}})] \cdot k_c}{T} \tag{66}$$

γ_{critic} denotes the critical deformation at the elastic-plastic transition, G' is the storage modulus from oscillation measurements, $\cos\delta^{-1}$ is derived from the loss factor $\tan\delta$, F_f is the fragmentation rate, $E_{\text{kin(b,a)}}$ represents the kinetic energy before and after collision, with $E_{\text{kin(a)}} = 0$ during agglomeration, E_{def} is the deformation energy of colliding particles, where k_c is a collision rate [67]. Now, let's further investigate the disturbances from forces and fluxes in the steady state.

$$\sum_i \delta X_i \cdot \delta J_i = -\frac{V}{\lambda \cdot T^2}\left[\left(\frac{\partial \tau_{(\text{oil})}}{\partial \varepsilon}\dot{\gamma} + \tau\frac{\partial \dot{\gamma}}{\partial \varepsilon}\right)\delta\varepsilon\right]^2 + \\ \frac{V \cdot G' \cdot \cos\delta^{-1}}{T}\left[2\gamma_{\text{critic}}\frac{\partial \gamma_{\text{critic}}}{\partial \varepsilon}\frac{\partial F_f}{\partial \varepsilon}(\delta\varepsilon)^2\right] + \frac{1}{T}\left[\frac{\partial E_{\text{kin(b)}}}{\partial \varepsilon}\frac{\partial k_c}{\partial \varepsilon}(\delta\varepsilon)^2\right] \tag{67}$$

The equation (67) consists of 3 terms, and we are seeking conditions for the right-hand side to become negative. This implies: - The first term is always negative due to the sign and the square. - For the second term, the condition is:

$$\frac{\partial \gamma_{\text{critic}}}{\partial \varepsilon} < 0; \ \frac{\partial F_f}{\partial \varepsilon} > 0 \quad \text{and vice versa.} \tag{68}$$

- In order for the third term to become negative, the following conditions must be met:

$$\frac{\partial E_{\text{kin(b)}}}{\partial \varepsilon} < 0, \ \frac{\partial k_c}{\partial \varepsilon} > 0 \quad \text{and vice versa.} \tag{69}$$

Of course, the highest probability exists when all 3 terms on the right-hand side are negative. In this context, the focus is on the second term. This means we are looking for a decreasing critical deformation γ_{critic} while the fragmentation rate F_f increases.

Now, let's inquire about experimental results reflecting such behavior (the second criterion). Rheometer studies in [72] are utilized for this purpose. There, oscillation measurements in amplitude-sweep mode were conducted using model greases. In this experimental procedure, the transition from predominantly elastic deformation to plastic deformation, quantified by γ_{critic}, can be observed. The influence of solid content on the critical deformation γ_{critic} appears to be clear. With increasing content, the critical deformation γ_{critic} decreases (Figure (26)). An explanation for this observation lies in the nature of the bonds present in the entire grease system. The physical bonds in the agglomerates are much weaker than the bonds in the base oil. This behavior is evident in rheometer experiments with two different types of solid materials. To incorporate the considerations in

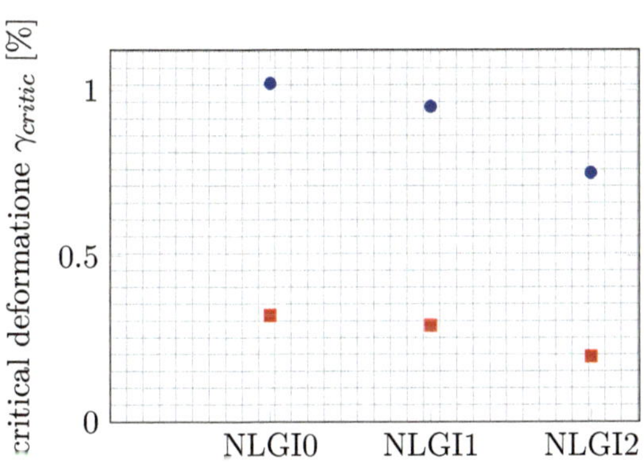

Figure 26: Critical deformation γ_{critic} vs. content of solid (NLGI-classes), PU-sample and Li-sample nach [72]

equation (67), shear experiments at constant shear rate were conducted using grease samples with varying solid content (Li-soap) in a rheometer. The experiments were performed using a plate-plate system and repeated 5 times.

Following these considerations, we can connect the attainment of the steady state with the possible occurrence of instabilities and the initiation of self-organizing processes (Figure (27)).

The previously seen correlation between the probability of structure formation and the solid content will now be investigated with an expansion of the considered dependencies. The starting point is the consideration of a general friction process of a viscoelastic lubricant film. The investigated dependencies are $\tau = \tau(\psi, \dot{\gamma})$ and $\lambda = \lambda(\psi)$. For the disturbance of the steady state, the following applies:

$$\frac{\partial}{\partial t}(\delta^2 S) = \delta X_B \cdot \delta J_B = \left(\frac{V}{T}\right)^2 \delta(\tau \cdot \dot{\gamma}) \cdot \delta\left(\frac{\tau \cdot \dot{\gamma}}{\lambda}\right) \tag{70}$$

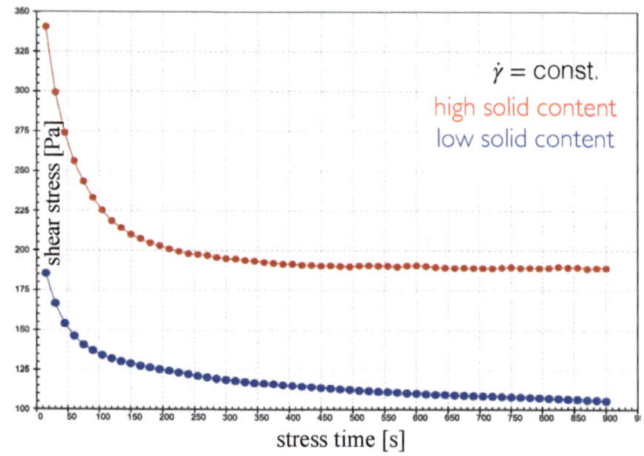

Figure 27: Temporal evolution of shear stress over time for two samples with different solid contents [67]

$$= \left(\frac{V}{T}\right)^2 \left[\frac{\dot{\gamma}^3}{\lambda^2}\frac{\partial \tau}{\partial \psi}\left(\lambda\frac{\partial \tau}{\partial \psi} - \tau\frac{\partial \lambda}{\partial \psi}\right)(\delta\psi)^2 + \frac{1}{\lambda}\left(\frac{\partial \tau}{\partial \dot{\gamma}}\dot{\gamma} + \tau\right)^2 (\delta\dot{\gamma})^2\right] \tag{71}$$

Only the pure terms are represented here, the mixed terms do not contribute significantly [73]. For the potential occurrence of self-organization, it is required that

$$\left(\frac{\partial \lambda}{\partial \psi}\right) > 0 \tag{72}$$

If ψ is associated with the solid content, then the condition found is quite realistic. This consideration also shows the influence of solid content on the possibility of triggering a self-organization process.

5.3.3 The Behavior of Grease under Triggered Self-Organization

This section begins with an experimental measurement showing the development of grease wear under frictional stress, following the described 3-step experimental procedure. The sample was subjected to different shear rates (shearing) in the rheometer, and the mechanical energy required to reach the crossover point was determined in an oscillation experiment (an indirect expression of grease wear).

The figure shows a decreasing grease wear with increasing shear rates up to $\dot{\gamma} = 2000\,[1/s]$. This observation is one we aim to associate with the phenomenon of self-organization. A description of this behavior concerning solid wear can be found, among other sources, in [57]. We intend to apply it to the issue of grease wear and expand it to include a consideration of temperature. Starting point is once again the entropy balance

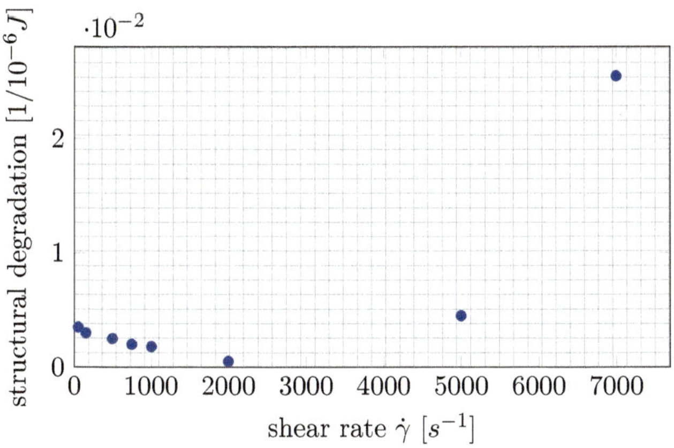

Figure 28: Grease wear of a PU model grease vs. Stress [74]

equation with

$$dS = dS_i + dS_e \tag{73}$$

The fact that a change in density distribution (solid distribution) occurs between differently stressed lubricating grease volumes (Figure (29)), leads to the consideration of two processes. For entropy production, it holds that

Figure 29: *left*: IR-topogramm of an unstressed grease sample, *right*: IR-topogram of the stressed grease sample [75]

$$\frac{dS_i}{dt} = X_1 \cdot J_1 + X_2 \cdot J_2 \tag{74}$$

and for the steady state and heat conduction, we have

$$X_1 = -\text{grad}\, 1/T \tag{75}$$

$$J_1 = -\lambda \operatorname{grad} T \tag{76}$$

Additionally, for a diffusion process [57]

$$X_2 = \operatorname{grad} \frac{\Phi}{T} \tag{77}$$

$$J_2 = -\gamma_D \operatorname{grad} \Phi \tag{78}$$

In this equation, T represents the temperature, λ denotes the thermal conductivity, Φ stands for the chemical potential, and γ_D is the transport coefficient. Following the assumption made in [57], a proportionality between wear and transport coefficient is described.

$$\frac{dS_i}{dt} = \frac{(\tau \cdot \dot{\gamma} \cdot V)^2}{\lambda \cdot T^2} + \gamma_D(T)(\operatorname{grad} \Phi)^2 T^{-1} \tag{79}$$

For the investigation in the steady state, a dependency $\gamma_D = \gamma_D(\tau \cdot \dot{\gamma} \cdot V)$ is assumed. Thus,

$$\frac{d}{d(\tau \cdot \dot{\gamma} \cdot V)} \frac{dS_i}{dt} = \frac{2(\tau \cdot \dot{\gamma} \cdot V)}{\lambda \cdot T^2} + \frac{\frac{d\gamma_D}{d(\tau \cdot \dot{\gamma} \cdot V)} (\operatorname{grad} \Phi)^2}{T} = 0 \tag{80}$$

$$\int \frac{2(\tau \cdot \dot{\gamma} \cdot V) \cdot d(\tau \cdot \dot{\gamma} \cdot V) \cdot T}{\lambda \cdot T^2 \cdot (\operatorname{grad} \Phi)^2} = \int -\gamma_D + c \tag{81}$$

and it follows that

$$\boxed{\gamma_D = \gamma_{D0} - \frac{(\tau \cdot \dot{\gamma} \cdot V)^2}{\lambda \cdot T \cdot (\operatorname{grad}\Phi)^2}} \tag{82}$$

as an analogous consideration to [57], with γ_{D0} as the integration constant. For the steady state, it is evident that with increasing stress on the grease, expressed by γ_D, the grease wear decreases. This representation describes the process depicted in Figure (28).

Now, let's observe the dependence $\gamma_D = \gamma_D(T)$.

$$\frac{d}{dT} \frac{dS_i}{dt} = \frac{-2(\tau \cdot \dot{\gamma} \cdot V)^2}{\lambda \cdot T^3} + (\operatorname{grad} \Phi)^2 \left[\frac{d\gamma_D(T)}{dT} T^{-1} - (T^{-2} \cdot \gamma_D(T) \right] = 0 \tag{83}$$

and assuming that

$$\frac{d\gamma_D(T)}{dT} \gg \frac{\gamma_D(T)}{T^2} \tag{84}$$

is valid. Further,

$$\int (\operatorname{grad} \Phi)^2 d\gamma_D(T) = \int \frac{2(\tau \cdot \dot{\gamma} \cdot V)^2}{\lambda \cdot T^2} dT \tag{85}$$

and

$$\gamma_D = -\frac{2(\tau \cdot \dot{\gamma} \cdot V)^2}{\lambda \cdot T (\operatorname{grad} \Phi)^2} + c \tag{86}$$

$$\boxed{\gamma_D = \gamma_{D0} - \frac{2(\tau \cdot \dot{\gamma} \cdot V)^2}{\lambda \cdot T(\mathrm{grad}\,\Phi)^2}} \tag{87}$$

With increasing temperature, the process approaches a limit. This limit could be close to the base oil properties, which, however, are never undershot [74].

5.3.4 Some Additional Insights into Possible Self-Organization Processes

Currently, there seems to be no experimentally feasible way to observe the formation of dissipative structures, i.e., self-organization, in the friction process of a grease film. The elegant example of Bénard cells, when applied to tribological contact, appears to be hardly observable. An experimental development in connection with rheometer tests is conceivable.

Thus, one must resort to indirect measurements or the description of theoretical indicators. Recording the wear of grease over stress is an example of this (Figure (30)).

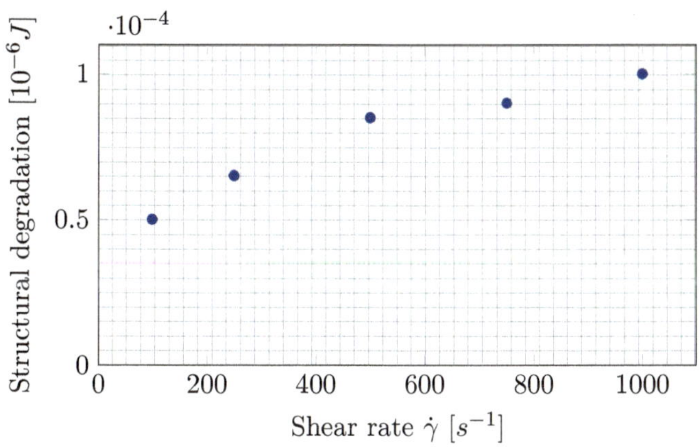

Figure 30: Increasing grease wear with increasing frictional load (from the 3-step experimental procedure)

The determination of process conditions to achieve an unstable state describes another possibility.

In extension of the considerations with Equation (67) and in reference to [35], let us examine the case where increasing deformation leads to an increase in frictional energy and, at the same time, grease wear also increases. This, in turn, results in a decrease in the required frictional energy. The aim is to observe whether considering these interactions also leads to greases with higher solid content being more likely to experience unstable states (see criterion (68)) and thus more likely to trigger self-organizing processes.

The interaction expressed for the frictional energy is given by

$$E_f = E_{f0} + E_{\dot{\gamma}} - E_{fw} \tag{88}$$

The first term represents the equilibrium state, the increase in frictional energy due to increased deformation is

$$E_{\dot{\gamma}} = (\tau \cdot \dot{\gamma} \cdot V) \quad \text{with} \quad \tau_{ostw} = k \cdot \dot{\gamma}^n \tag{89}$$

and as a general approach, we write

$$E_{fw} = b \cdot W_{ear}^q \tag{90}$$

Now, Equation (88) becomes

$$E_f = E_{f0} + (\tau_{ostw} \cdot \dot{\gamma} \cdot V) - b\left(\frac{E_{f\dot{\gamma}}}{E_P}\right) \tag{91}$$

where $W_{ear} = E_{f\dot{\gamma}}/E_P$ from [65] with the interpretation that $E_P = T/B$. Here, B is the degradation coefficient according to BRYANT ET AL. [48].

Now we investigate the disturbance from the steady state with Equation (62) which reads

$$\frac{1}{2}\frac{\partial}{\partial t}(\delta^2 S) = \delta\left[\frac{k\dot{\gamma}^n\dot{\gamma}m}{\rho} - b\left(\frac{k\dot{\gamma}^n\dot{\gamma}m}{\rho E_P}\right)^q\right]\delta\left[-\frac{k\dot{\gamma}^n\dot{\gamma}m}{\rho\lambda T^2} + \frac{b}{\lambda T^2}\left(\frac{k\dot{\gamma}^n\dot{\gamma}m}{\rho E_P}\right)\right] \tag{92}$$

We introduce a parameter ε that describes the distance or variation from the equilibrium state. We observe the dependencies $\rho(\varepsilon)$ and $\lambda(\varepsilon)$. For better readability, we write

$$\frac{1}{2}\frac{\partial}{\partial t}(\delta^2 S) = -\frac{1}{T^2}\left[\text{Factor}_1\left(\frac{\partial\rho}{\partial\varepsilon}\right) + \frac{1}{\lambda^2}(\text{Factor}_2)\frac{\partial\rho}{\partial\varepsilon}\frac{\partial\lambda}{\partial\varepsilon}\right](\delta\varepsilon)^2 \tag{93}$$

with Factor_1

$$\left(-\frac{a}{\rho^2} + \frac{qba^q}{E_P^q}\cdot\frac{1}{\rho^{q+1}}\right)^2 \tag{94}$$

and Factor_2

$$\left(\frac{a^2}{\rho^4} - \frac{ba^{q+1}}{E_P^q\rho^{q+2}} - \frac{qba^{q+1}}{E_P^q\rho^{q+3}} + \frac{qb^2a^{2q}}{E_P^{2q}\rho^{2q+1}}\right) \tag{95}$$

We set $a = k \cdot \dot{\gamma}^{n+1} \cdot m$. Experimental investigations on b and q could not be conducted. Therefore, it is not possible to estimate the factors. However, let us return to the research objective with the question: does the condition of increasing solid content also appear favorable for potential self-organization here? One of several possibilities for the right-hand side of the equation to become negative is the condition that

$$\frac{\partial\rho}{\partial\varepsilon}\frac{\partial\lambda}{\partial\varepsilon} > 0 \tag{96}$$

In interpretation, this condition again describes the influence of increasing solid content on possible structure formation. However, until experimental results are available, this circumstance appears as one of several possible factors.

Another indication of indicators for the formation of dissipative structures is given by PRIGOGINE [2] and KLAMECKI [35]. They point to the cyclic behavior of energy dissipation rates . For the friction process in the grease, the concepts of B.KLAMECKI are modified and adapted. A stressed grease volume element and various dissipation mechanisms are considered. Selected are

- thermal dissipation E_k
- mechanical dissipation through the fragmentation process E_i
- mechanical dissipation through coagulation after collisions of solid particles E_j

The relative energy dissipation of one mechanism in relation to the relative energy dissipations of the other mechanisms will be investigated. The following assumptions about the influence on the energy dissipation rate are made:

- mechanical energy dissipation always occurs with thermal dissipation
- the size of the energy with which the system is affected, $E_0 = E_i + E_j + E_k$, has an influence
- the dissipation that has occurred so far has an influence
- the relative dissipation of the mechanisms in relation to the relative energy dissipation of the investigated mechanism

It is now generally written

$$\frac{dE_i}{dt} = zE_iE_k + a_iE_0 - b_iE_i^m + cE_kE_iE_j \tag{97}$$

For the interaction of the mechanisms, let

$$c_i = \left[1 - \frac{d_{ij}E_j^{pi} + d_{ik}E_k^{pk}}{d_{ii}E_i^{pi}}\right] \tag{98}$$

and for the other interactions it holds

$$c_j = \left[1 - \frac{d_{ji}E_j^{pj} + d_{jk}E_k^{pk}}{d_{jj}E_i^{pj}}\right] \tag{99}$$

$$c_k = \left[1 - \frac{d_{ki}E_j^{pk} + d_{kj}E_k^{pk}}{d_{kk}E_i^{pk}}\right] \tag{100}$$

and further

$$z_i = \left[1 - \frac{d_{ik}E_k^{pk}}{d_{ii}E_i^{pi}}\right] \tag{101}$$

$$z_j = \left[1 - \frac{d_{jk}E_k^{pk}}{d_{jj}E_i^{pj}}\right] \tag{102}$$

Equation (97) can now be rewritten as

$$\frac{dE_i}{dt} = \left[1 - \frac{d_{ik}E_k^{pk}}{d_{ii}E_i^{pi}}\right] E_iE_k + a_iE_0 - b_iE_i^m + \left[1 - \left(\frac{d_{ij}E_j^{pi} + d_{ik}E_k^{pk}}{d_{ii}E_i^{pi}}\right)\right] E_iE_jE_k \tag{103}$$

The parameters z, a, b, and c are general functions of the considered energy dissipation mechanisms. The parameters d and p describe the individual material behavior. Experiments to determine these parameters are not yet possible, so numerical evaluation is attempted to understand the principal behavior. This involves answering the question: Are there conditions that lead to a cyclical behavior of the relative energy dissipation rate? A cyclical behavior, according to [24] and [34], indicates the possibility of forming dissipative structures.

It is now desired, for better visualization, to span a two-dimensional surface, where the ordinate, for example, represents a mechanical dissipation rate and the abscissa describes a thermal dissipation rate. By comparing the slopes and generating a direction field, to ultimately depict the trajectories, the development of relative energy dissipation rates becomes evident. The numerical calculations were performed using *Geogebra* at the University of Vienna.

Figures (31) and 32) show that there are conditions under which the model used leads to interesting, even cyclical developments in the relative energy dissipation rate. So far, the parameters in the model have been chosen arbitrarily and the axis labels are only exemplary; experimental investigations are now required to quantify these parameters.

This investigation supports the speculation in [74] and [67] that the formation of dissipative structures is also possible in stressed grease films.

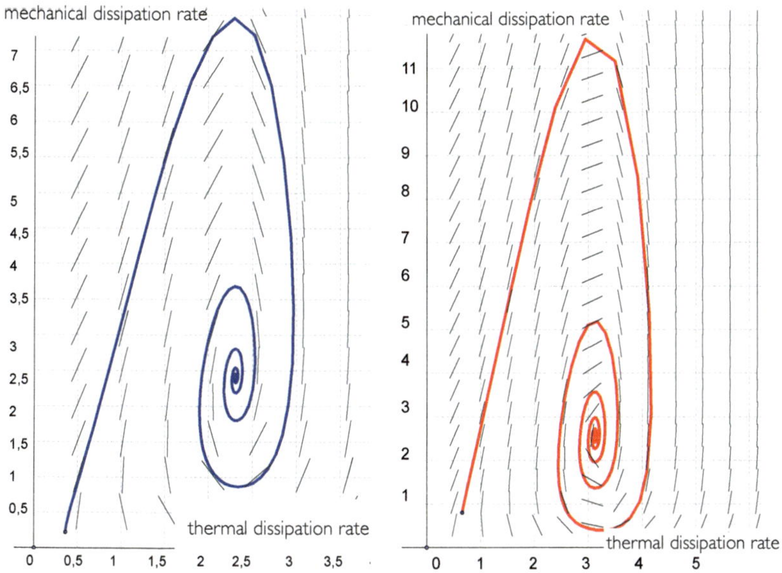

Figure 31: Two examples of cyclical behavior of relative energy dissipation rates

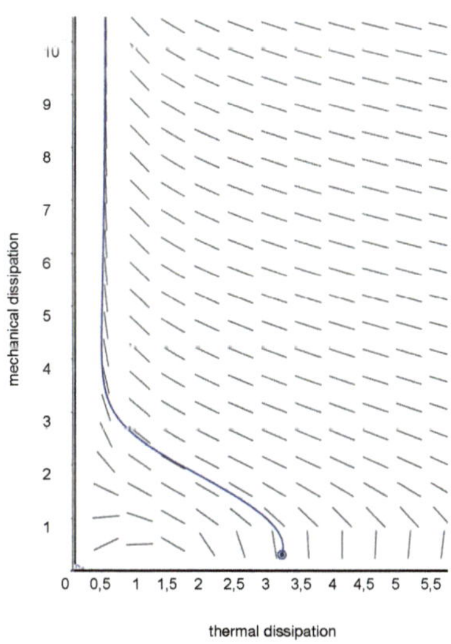

Figure 32: Example of the transition from dominant thermal dissipation to dominant mechanical dissipation

6 Lubricating Grease

6.1 History and Descriptions

6.1.1 Historical Background

The initial applications of grease in technological development are difficult to date precisely. It is presumed that lubricants were already used around 2400 BC to reduce friction during the transportation of large stone blocks. From the beginning of the Common Era, Greek and Roman chariots (see Figure (33)) are known, requiring lubrication for their wheel bearings. These lubricants, whether oils or fats, were likely of animal or plant origin, such as olive oils or sheep and beef tallow [76]. Oils and more consistent substances were probably used alongside each other and further developed. For instance, the axles of Egyptian chariots were lubricated with grease around 1400 BC. A sample from the Cairo Museum surprisingly indicated not only a plant-based or animal-based fat, but also a lime soap [76]. In the southern Spanish university city of Huelva (Andalusia), there is a wheel with a diameter of approximately 6 meters from the 2nd century, which was used for dewatering in underground iron ore mining (see Figures 34 and 35). The hoisting wheels were mounted on sliding bearings (metal/wood) and were likely lubricated with water or a visco-elastic lubricant similar to the described wheel hubs. An interesting aspect is that the direct bearing location in the wooden structure was replaceable. This indicates a deliberate focus on directing the solid wear to the softer friction partner, allowing for the convenient replacement of the worn part.

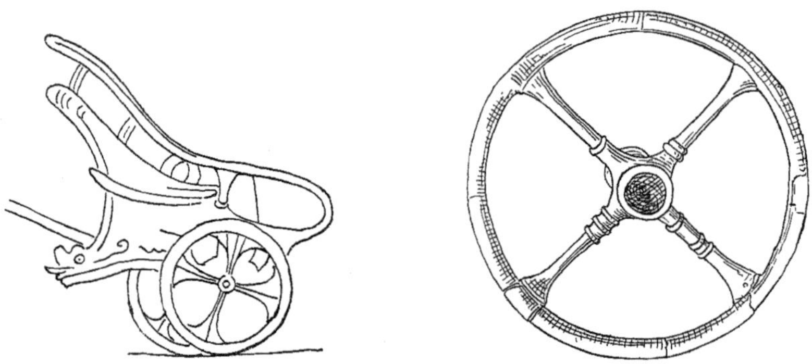

Figure 33: Greek chariot and bronze wheel (Beginning of the Common Era) [77]

An interesting study on the history of lubrication can be found in [78]. It shows that in the 17th century, wood tar was used for wagon lubrication. The authors interpret a

Figure 34: Arrangement of hoisting wheels for dewatering in mining in the 2nd century, model from Museo De Huelva, Andalusia

Figure 35: Historical bearing of a hoisting wheel and metal axles (Museo De Huelva, Andalusia)

container depicted in an engraving by Lucas Cranach the Elder at the rear of a wooden wagon as a lubricant container. It was used to carry the necessary means for lubricating the wagon axles (see Figure 36). This was a type of lubricant that, besides its main component of wood tar, included additives such as pine oil, linseed oil, beeswax, and likely spoiled animal fat (Wolf, P.: The land and people of the central Eastern March (in German), among others mentioned in [78]).

With the onset of the industrial age, the development of lubricants accelerated rapidly. Modern grease development progressed through sodium greases (1872), calcium and aluminum greases (1882), lithium greases (1942), and calcium complex soap greases (1940). In 1952, the first aluminum, barium, and lithium complex soap greases were patented [79].

Figure 36: Wagon with grease pot (detail by Lucas Cranach the Elder) [78]

6.1.2 On the Definition of Lubricating Greases

With the technical development of lubricants and the equally rapid advancement of the friction pairings to be lubricated, new perspectives have emerged in the description of lubricating greases. However, the fundamental characterization has not changed. Nevertheless, entirely new developments suggest imminent innovations.

Here are some examples of definitions or descriptions of lubricating grease :

- Lubricating greases are consistent lubricants composed of mineral oil and/or synthetic oil and a thickening agent. They may contain additives and/or solid lubricants [80].

- Lubricating grease is a multiphase material consisting of 70-90% base oil, 3-30% thickening agents, and additives [81].
- A lubricating grease is a solid or semi-fluid substance resulting from a dispersion of a thickening agent in a liquid lubricant; other components that impart specific properties may be included [82].
- Lubricating greases are ointment-like, plastically deformable substances in which oils are incorporated into a metal soap framework or swollen thickening agents. These oils provide the actual lubrication. The thickeners or gelling agents also largely support lubrication [83].
- Lubricating grease is a system composed of a solid thickening agent and a liquid base oil [84].
- Lubricating grease is a consistent lubricant, considered a mixture of a heterogeneous solid phase with a heterogeneous liquid phase [85].
- Grease is a lubricant that, under certain loads and within its temperature range, exhibits the properties of a solid, becomes plastically deformed upon reaching the critical point, begins to flow like a liquid, and reverts to solid properties once the load is removed [86].
- Lubricating greases are semi-fluid lubricants composed of a base oil thickened with 5 to 30 percent by weight of various thickening agents [87].
- Lubricating greases are consistent lubricants that represent dispersions of a thickening agent in a liquid lubricant [88].
- Lubricating greases are three-dimensional networks consisting of a thickening agent and a base oil [89].
- Lubricating greases are non-stationary physical systems [90].
- Lubricating greases are colloid-disperse systems and exhibit a yield point [91].
- Lubricating greases are colloidal dispersions, more precisely, suspensions [92].
- Lubricating grease = base oil + thickening agent + (additives) [93].
- Lubricating greases are physically considered suspensions of a solid and a liquid phase [94].

- Lubricating grease is defined as a solid to semi-fluid product or a dispersion of a thickening agent in a liquid lubricant [95] cited in [81].
- Lubricating greases are mixtures of a liquid and a solid phase. The liquid phase is a base oil, e.g., mineral oil and/or synthetic oil, and the solid phase is a thickening agent, e.g., a soap, with additives (additives and solid lubricants) included to enhance desired properties [96].
- A lubricating grease mixes and disperses lubricating oils in a thickening agent to form a gel-like product [53].
- A lubricating grease is a multiphase system consisting of three components: thickener, base oil, and additives [97].
- A lubricating grease is a polydisperse system in which one phase (the dispersed phase) is dispersed in another phase (the dispersing medium). The system is chemically and physically heterogeneous. The lubricating grease can be considered a macroscopically dispersed heterogeneous mixture [98].
- Lubricating greases are generally highly structured suspensions made from thickening agents dispersed in mineral oil or synthetic oil [99].

This list could be easily extended and shows that, although the definitions have similarities, there are different approaches to the study of the lubricating grease problem. A description derived from the mentioned definitions and the presented studies is already given by the definition (2.14) and reads as follows:
Lubricating greases are colloid-dispersed systems. They consist of a base oil and a solid. Lubricating greases possess pronounced visco-elastic properties.

6.2 Selected Types of Lubricating Grease

A classification of lubricating greases into metal soap greases and non-soap greases, as is usually done, already indicates the solid used. For practical application, simple metal soaps, complex metal soaps, or mixed soaps used as solids are of particular relevance. Non-soap greases are used much less frequently. For practical use, the preferred application is often of greater interest, and practitioners have adopted classifications such as chassis grease, gear grease, roller bearing grease, etc., or based on properties like fluid grease, block grease, high-temperature grease, etc.

In development are biogenic lubricating greases. This means that, in addition to the base oil, the solid (thickener) also exhibits biogenic properties [100], [101], [102].

6.2.1 Thickener Types (Solids for Common Lubricating Greases)

The special properties of lubricating greases, differentiated from the base oil, are based on the solid used. Metal soap greases (which are the subject of the present study) contain metal soaps, i.e., salts of fatty acids with the oxides or hydroxides (Schmidt, G. - Chemistry and Production of Lubricating Greases using Metal Soaps as an Example. in [92]) of the following metals:

- Lithium
- Sodium
- Calcium
- Aluminum

The metal-free organic part of the metal soap thickeners contains at least one fatty acid residue, which forms the hydrocarbon skeleton of the thickener molecules. Also in (Schmidt, G. - Chemistry and Production of Lubricating Greases using Metal Soaps as an Example. in) [92], you can find a detailed list of possible fatty acids, among which the 12-hydroxystearic acid obtained by hydrogenation of castor oil occupies a special position. Because its Li, Na, and Ca soaps form particularly stable multipurpose greases. In ad-

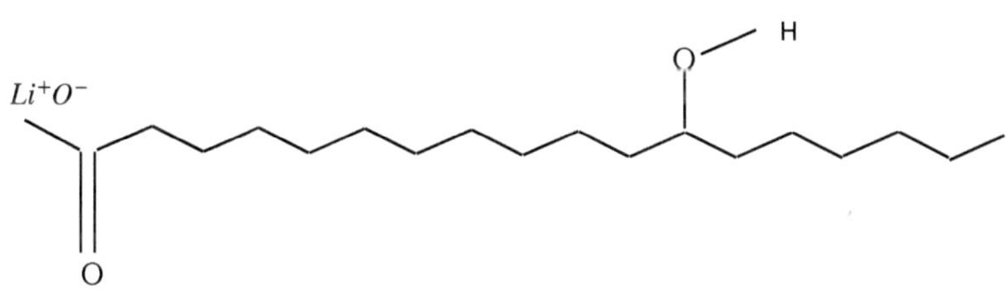

Figure 37: Structural formula for lithium 12-hydroxystearate

dition to simple soap greases, which are the most commonly used, there is the group of complex soap greases. In [96], we see:

- Calcium complex
- Lithium complex
- Aluminum complex
- Calcium sulphonate complex

- Other, e.g., sodium terephthalate

They are characterized by special performance. Complex soaps are composed of a base, a fatty acid, and a non-fatty acid. Such non-fatty acids can be, for example, acetic acid, boric acid, etc. As a result, there is a complex mixture with various thickener molecules (Schmidt, G.: Chemistry and Production of Lubricating Greases using Metal Soaps as an Example, in [92]), [103]. Other possible thickeners for the production of lubricating greases, in addition to the mentioned soaps and complex soaps, can include modified layered silicates, polyureas, carbon black, colloidal silica, pigments, PTFE, and others [104]. Among them, organophilic bentonites and polyureas certainly have the greatest significance, as their lubricating greases are particularly suitable for high temperatures. In [96], non-soap solids are listed as:

- Polyurea (organic)
- Bentonite/Clay
- Silica gel/HDK (inorganic)
- PTFE
- Others

Important properties of the solids used include the dropping point, water resistance, corrosion protection, EP properties, and others. Information can be found in [92], [96], [81].

6.2.2 Base Oils

Due to their large proportion in the percentage composition of lubricating greases, the base oils used play a prominent role. They significantly influence the viscosity-temperature behavior, cold behavior, or suitability for sealing materials.

Different types of base oils are generally distinguished [96][104]:

- *Mineral oils*, which are obtained from petroleum vacuum distillation and subsequent refining
- *Synthetic oils*, e.g., diesters, polyglycol ethers, silicone oils, polyether oils (PFPE) [96][105].

Naturally, *natural oils* have also been included for many years. Rapeseed oil has emerged as a base oil for lubricants, and rapeseed oil-metal soap greases can also be produced with it.

Mineral Oils The most common types of mineral base oils, according to [96][104], are formed by:

- *Distillates, Refined Distillates*: used for minor application issues
- *Acid-treated, Solvent-treated Raffinates*: obtained from refined distillates and used in the majority of lubricating greases
- *Bright Stocks*: used as base oil for high-temperature greases after deasphalting and dewaxing

Base oils are also classified according to the nature of the crude oil, i.e., *aromatic, naphthenic*, and *paraffinic*. The importance of aromatic oils is decreasing, while paraffinic oils are increasingly used in lubricating grease production [106]. Naphthenic mineral oils are used for the production of oils with low or high viscosity index in two different manufacturing processes [104].

Generally, for the production of lubricating greases, mineral oils with kinematic viscosities ranging from $\nu = 40mm^2/s$ to $\nu = 3400mm^2/s$ (at $20°C$) are used. Lower viscosities are employed for high relative velocities, while higher viscosities indicate higher loads and lower relative velocities.

Synthetic Oils The proportion of lubricating greases with synthetic base oils is comparatively low. This is also due to their relatively high price, so their use is only justified when the properties of mineral oil greases reach their limits. The use of *polyalphaolefins, esters,* and *alkylbenzenes* has proven effective. However, *silicones* or *polyphenyl ethers* are also successfully applied. For very wide temperature ranges, base oils with very low pour points and weaker temperature-viscosity behavior (at higher temperatures) are preferably used for grease production. For example, diesters, silicone oils, and polyalphaolefins are utilized. A comprehensive overview of synthetic lubricating greases is provided in [92], [107], and [96].

Biodegradable Base Oils/Lubricating Greases Biodegradable base oils have been in use since the 1970s. Their development has progressed rapidly, and the range of products available for use has expanded enormously. The basis includes [108]:

- Native oils, e.g., rapeseed oil
- Synthetic esters

Lubricating greases based on rapeseed oil or synthetic esters are biodegradable. Guidelines in this regard include:

- The content of renewable raw materials according to ASTM D-6866 must be at least 25%.

- Biodegradability according to OECD 301 B must be at least 50%.
- The lubricating grease must not be labeled as environmentally hazardous [109].

In an experimental study [110], M. Fiedler et al. investigate the influence of base oil polarity on solid wear in grease-lubricated pairs. Model greases with biodegradable base oils are used. The study compares *PAO - synthetic reference oil, HOSO - high oleic sunflower oil, TMPO - trimethylpropane trioleate*, and *OCT - octyldodecyl isostearate.*

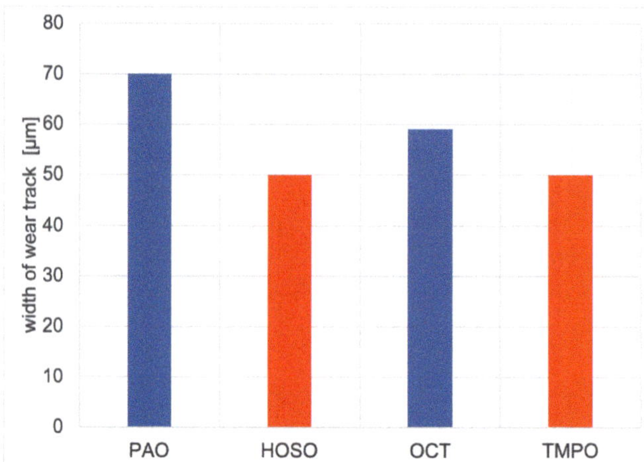

Figure 38: Disk wear using different base oils according to [110] in ball-disk contact

Considering that polarity shows differences with OCT $\approx$ PAO $\ll$ HOSO $\approx$ TMPO [111], the influence of base oil polarity on measured solid wear is noticeable. In tribometer investigations in ball-disk contact, high polarities show a favorable effect. Detailed information can be found in [111] and [110].

6.2.3 Biogenic Lubricating Greases

The term *biogenic* can describe a material that is of biological or organic origin. In contrast to traditional lubricating greases, as well as biodegradable lubricating greases, a *biogenic* lubricating grease consists of a biological base oil such as rapeseed oil, sunflower oil, etc., and a biological solid (thickener) such as cellulose, beeswax, etc. Various research groups are advancing the development of usable *biogenic lubricating greases.* These research efforts are particularly urgent not only for pragmatic considerations regarding the finite nature of resources such as lithium, aluminum, calcium, etc., but also due to considerations of sustainability and environmental compatibility.

Initial research on fully biogenic lubricating greases was conducted by the research group of J.M. Franco [112], [113], [114], [115], [116], [117], [118], [119], [120], where castor oil or soybean oil was used as biodegradable lubricating oil and conventional thickeners were replaced by natural thickeners such as cellulose derivatives, chitin, chitosan, glycerol stearates, or sorbitan stearates. These comparative studies of biobased fats with other conventional fats such as lithium-12-hydroxystearate grease show that some of the biobased fats exhibit comparable thermal, mechanical, and rheological properties for potential lubrication applications [121]. An extensive investigation was also conducted in [122] by N. Acar. There, sunflower oil and castor oil were used as base oils. Biogenic solids included lignin/polyethylene glycol diglycidyl ether, chitosan, ethyl cellulose, natural cellulose, natural wood cellulose, beeswax, or straw, among others.

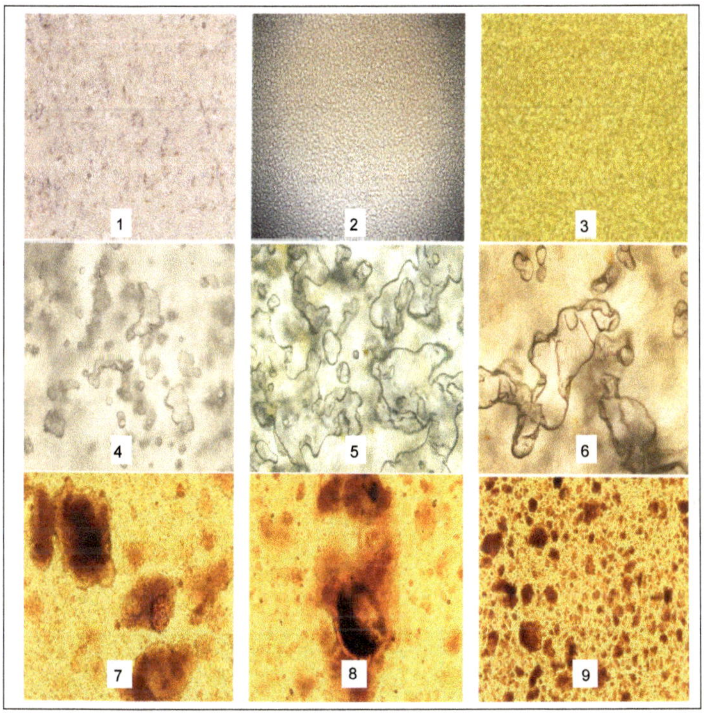

Figure 39: Transmitted light microscopy images of selected biogenic grease samples [123]

In Figure (39), the labels denote:

1. *HOSO and rapeseed oil with beeswax, glyceryl monostearate, and cetyl alcohol,*
2. *HOSO and glycerol with cellulose ether*

3. *HOSO with lignosulfonates*
4. *HOSO with natural cellulose fibers, 18µm*
5. *HOSO with natural cellulose, 20-40µm*
6. *HOSO with natural wood pulp from softwood, 70-150µm*
7. *Rapeseed oil with lignin/PEGDGE (weight ratio of 1/0.25)*
8. *Rapeseed oil with lignin/PEGDGE (weight ratio of 1/1)*
9. *Rapeseed oil with lignin/HMDI (weight ratio of 1/2)*

Here, *HOSO* stands for high-oleic sunflower oil, *PEGDGE* for polyethylene glycol diglycidyl ether, and *HMDI* for hexymethylen. In this microscopy procedure, a window size of *683.5 µm* x *536 µm* was chosen for each image to display the different structures.

In investigations on a ball-on-disc tribometer in steel-steel contact (100Cr6 for the ball with a diameter of 12.7mm and S235JR for the disc) and a Hertzian pressure of 0.93 GPa, the results depicted in Figure (40) were obtained for a sliding speed of 0.129m/s.

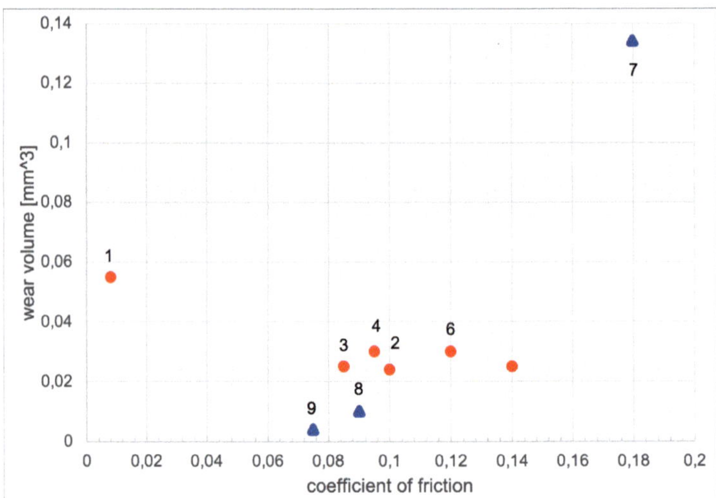

Figure 40: Friction and wear behavior of the model greases shown in Figure (39) [121]

The results presented are to be regarded as the first study of model substances aimed at leading to usable biogenic greases. According to their composition, they are divided into red and blue categories.

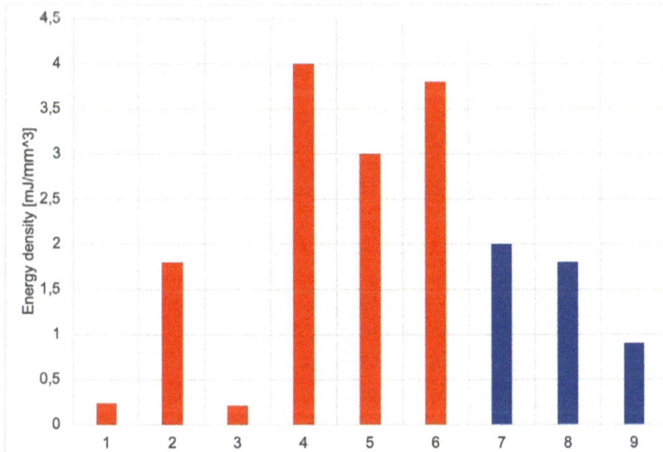

Figure 41: Applied energy densities during shearing at $\dot{\gamma} = 1s^{-1}$, $\vartheta = 25^0C$, and a shear time of $t = 3600s$.

Friction in the grease film can be comparatively described by determining the energy density e_{rheo} required during shearing in the rheometer (Figure (41)). In all tests, the constant shear rate $\dot{\gamma}$, test temperature ϑ, and test time t are the same. There are significant differences, primarily attributable to the very different structures. Samples 4, 5, and 6 exhibit remarkably high applied energy densities. These materials contain cellulose fibers with large lengths.

6.2.4 Grease Structure

For the investigations presented in this book, the term *grease structure* plays a significant role. The following description and definition are proposed to enable a unified understanding .

Definition 6.1 (Grease Structure) *The grease structure describes the geometry, arrangement, and distribution of the solid material (thickener) in the base oil of a grease.*

This structure forms differently depending on the base oil used, the type of solid material, the solid content, and specific manufacturing conditions.

Various experimental methods are available for the investigation and representation of grease structure. To visualize geometry and distribution, AFM (Atomic Force Microscopy), SEM (Scanning Electron Microscopy), IFM (Interferometry), and LM (Light Microscopy, reflected/transmitted) are used.

Early detailed illustrations of selected grease structures can be found in [124]. Around the same time, [125] describes his ideas on the positioning of base oil in grease structure, including:

- through intermolecular attraction between soap and the polar components of the oil
- through capillary effects
- through "mechanical"retention

Similar descriptions are provided by MANG and DRESEL [126], referring to [127].

In investigations at the macroscopic scale, highly differentiated distributions of agglomerates are observed. Transitioning to the microscopic level reveals differences in the geometry of the structure-forming solid elements.

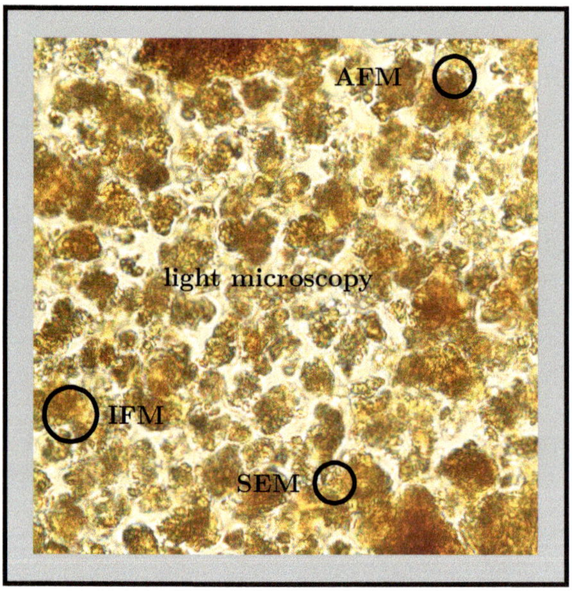

Figure 42: Different investigation methods for structure representation: Bright-field/Transmitted Light Microscopy, e.g., window size $500\mu m$; Atomic Force Microscopy, e.g., $20 \times 20\mu m$; Scanning Electron Microscopy, e.g., $30 \times 30\mu m$; and Interferometry, e.g., $40 \times 40\mu m$.

The choice of investigation method also depends on the objectives of the study. Agglomerate distributions and arrangements are most effectively studied at the macro scale, while other subjects of investigation are better studied at the micro scale. Figure (42) illustrates this comparison, with the provided resolutions being exemplary values (see also [128]).

Examples of structures illustrating different solid particles and investigation methods will now be presented.

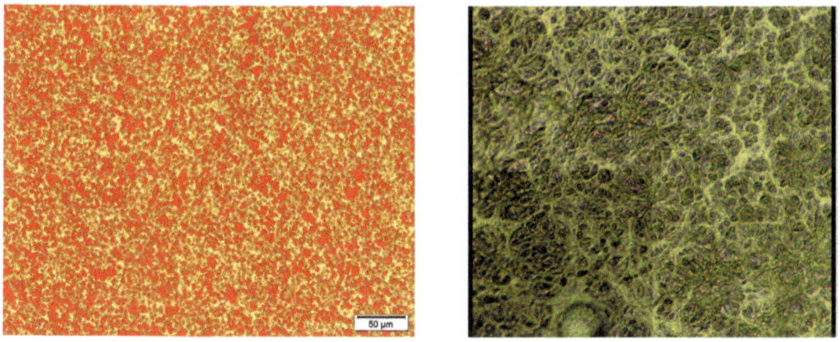

Figure 43: *Left*: Transmitted Light Microscopy *Right*: Brightfield Microscopy (Window 210μm) for Distribution and Structure Formation

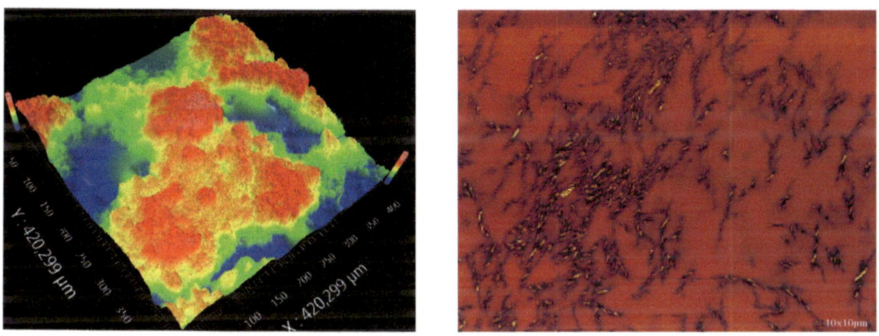

Figure 44: *Left*: Interferometry of the Solid *Right*: Atomic Force Microscopy of the Solid in the Base Oil

The sample material was provided by Fuchs Europe Schmierstoffe (Mannheim, Germany) and Fuchs Lubritech (Kaiserslautern, Germany). First, several illustrations of different investigation methods (Figures 43, 44, 45).

The application of SEM, AFM, IFM, or LM (light microscopy) not only differs in the measurement method but, crucially for tribology of grease, in sample preparation. The significant advantage of atomic force microscopy, unlike SEM, IFM, and sometimes LM, is the absence of separation between base oil and solid particles. The substantial manipulation of samples caused by the separation of base oil and solid particles deviates significantly from

Figure 45: Example of solid particle investigation using SEM

the original structures in tribological contact, which is undesirable. Important insights are provided by the research group at the University of Huelva in [129].

Of course, sample material must also be prepared to some extent for AFM examination. Whether by smoothing or heating, a flat grease film must be produced.

Images from SEM examinations of vastly different solid particle structures are shown in the following figures. Biogenic grease samples were also used in these examinations.

The results of such investigations are impressive for comparing the unstressed and stressed lubricant structures. Here, there can be a very clear difference in structure. Such *before* - *after* representations can provide a qualitative impression of lubricant wear (Figure 48, also see [128]).

The conditions of the manufacturing process also have an influence on structure formation [130], but these will not be discussed here.

6.3 Selected Test and Experimental Facilities

Under *Testing facilities*, experimental facilities for standardized testing of selected characteristics, properties, or behaviors of lubricating greases are understood. In contrast, *experimental facilities* are experimental setups for investigating specific behaviors, properties, etc. with non-standardized procedures.

Numerous references to testing facilities can be found in [92] and [81], including:

- FE8 - Rolling bearing test *DIN 51819*
- Dropping point according to *DIN ISO 2176*
- Cone penetration according to *DIN 150 2137*
- Penetration walk stability according to *DIN ISO 2137*

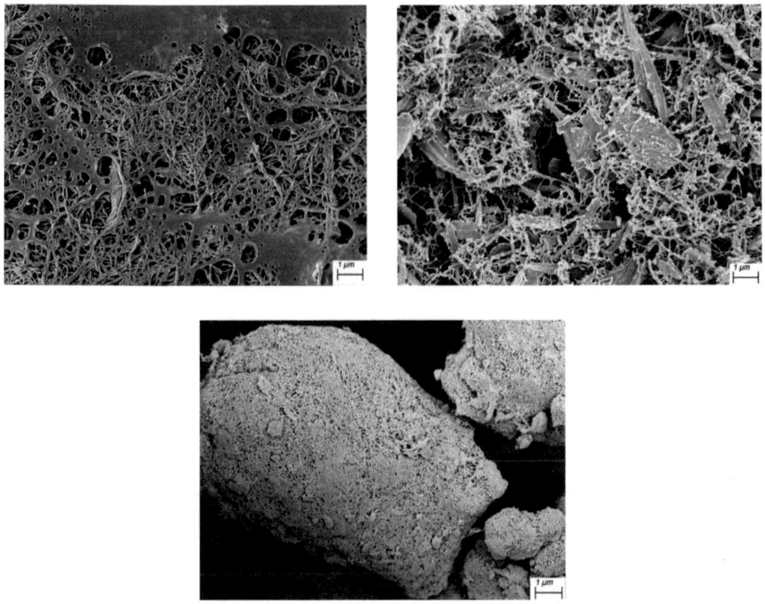

Figure 46: SEM images of 3 biogenic solids of model greases. From left: Cellulose in glycerol, glyceryl monooleate; Polyhydroxybutyrate, Ethyl cellulose in MTC oil, castor oil; Beeswax, glyceryl monostearate, cetyl alcohol in HOSO (high oleic sunflower oil), castor oil

- Lubricating grease service life FAG FE9 according to *DIN 51821*
- Flow pressure according to *DIN 51805*
- Corrosion protection properties Emcor test according to *DIN 51802, ISO 11007*
- TIMKEN test according to *ASTM-D 2509*
- Shell Four-Ball Tester according to *DIN 51350 1-5; ASTM-D 2266*
- Water washout test according to *DIN 51807; ASTM-D 1264*
- Oscillating friction wear device according to *DIN E 51834*
- FZG gear rig according to *DIN 51354 1/2*
- Roll stability tester according to *ASTM-D 1831*

Some examination possibilities are explicitly highlighted here.

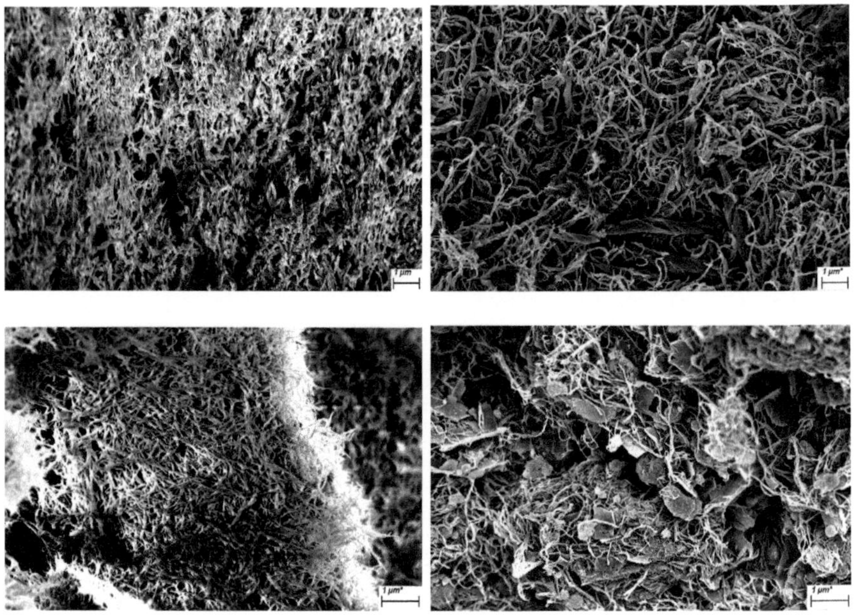

Figure 47: SEM images of solids, top: Li-12-hydroxystearate in castor oil; Li-12-hydroxystearate in PAO, bottom: Ca-12 hydroxystearate in PAO; Polyurea in PAO

Cone Penetration (DIN 150 2137) The penetration test is used to determine the consistency of a lubricating grease. It measures the depth of penetration established after 5 seconds by a standardized cone (loaded by the weight of the cone mass) into a prepared grease sample (see Figure 49). A high penetration depth (in 1/10 mm) describes a *soft* grease, and vice versa. To describe the consistency of stressed lubricating greases, the so-called *walk penetration* is measured. For this purpose, the grease sample is filled into a grease worker (Figure (50)) and subjected to a predetermined number of double strokes (60, 5000, 60000, 100000). Subsequently, the penetration decrease is measured.

The measured penetration depths were assigned consistency grades and published by the National Lubricating Grease Institute as NLGI grades.

Timken Test Setup The lubricant under test is examined in a friction system consisting of a rectangular test block, through which the test forces are applied, and a rotating test cup (see Figure 51). The test force is changed step by step. The result determines a pass load and a seizure load. Wear is determined by individual weighing of the two test specimens. The test time per force step is 10 minutes.

Figure 48: *Left:* Top - unstressed structure of a polyurea solid, bottom - cellulose, *Right:* Top - frictionally stressed PU structure, bottom - cellulose

FE 8 Rolling Bearing Test Rig This rig investigates bearing- and lubricant-specific influences on the wear and friction behavior of rolling bearings.
Speed: continuously adjustable from 7.5 to 3,000 rpm
Standard: 7.5, 75, 750, 1,500, 3,000 rpm
Axial Load: 100 to 100,000 N; controllable by force measurement cell
Temperature: 20 – 100°C
Friction Conditions: boundary and mixed friction, hydrodynamics
Lubricants: grease; oil circulation lubrication
Measured Variables: continuous friction torque measurement, vibration measurement
Test Bearings: angular contact ball bearings 7312B, tapered roller bearing 31312A, ball bearing 6312, axial cylindrical roller bearing 81206, axial cylindrical roller bearing 81212
Test standards include:

- DIN 51819: Mechanical-dynamic testing on the rolling bearing lubricant test rig FE8 – Part 2: Procedure for greases, test bearings to be used, angular contact ball bearings or tapered roller bearings

Figure 49: Micro-penetrometer with measuring cup

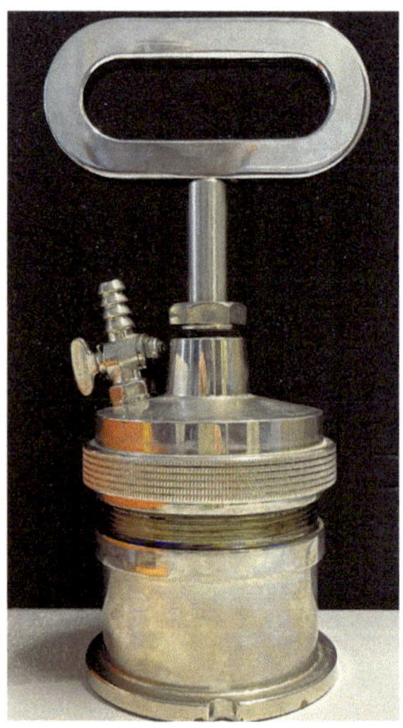

Figure 50: Manually grease worker

NLGI Grade	Worked Penetration [0.1 mm]	Description	Application
000	445-475	Very soft	Gears
00	400-430	Fluid	
0	355-385	Still fluid	Gears
1	310-340	soft	Bearings
2	265-295	Greasy	Bearings
3	220-250	Transition to firm	Water pump grease
4	175-205	Firm	
5	130-160	Very firm	Block grease
6	85-115		

Table 6.1: NLGI Grades and Penetration

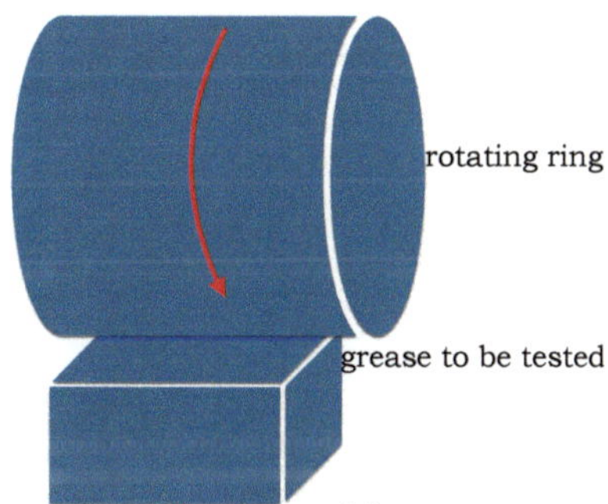

Figure 51: Timken Test Setup

- DIN 51819: Mechanical-dynamic testing on the rolling bearing lubricant test rig FE8 – Part 3: Procedure for lubricating oil, test bearings to be used, axial cylindrical roller bearings
- Lubricant test rig FE8 for preselection and testing of greases and oils; part of EN 12081.

An interesting variation of the FE8 testing device is implemented by [132]. They investigate the wear behavior of grease-lubricated rolling bearings under oscillating motions (see Figure 53).

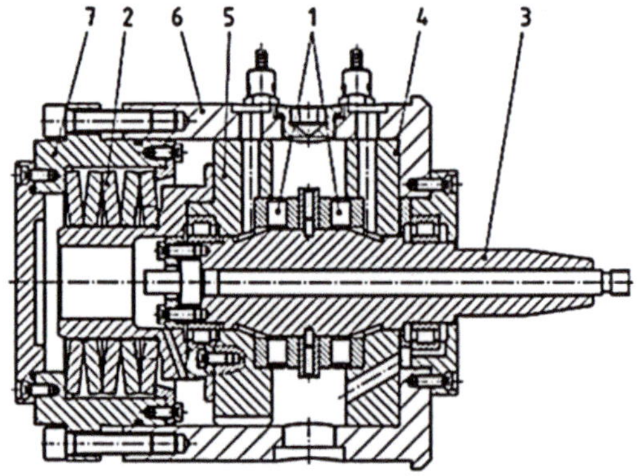

Figure 52: FE8 test device from [131], 1—test bearing, 2—spring package, 3—shaft, 4—bearing housing drive side, 5—bearing housing spring side, 6—housing, 7—cup.

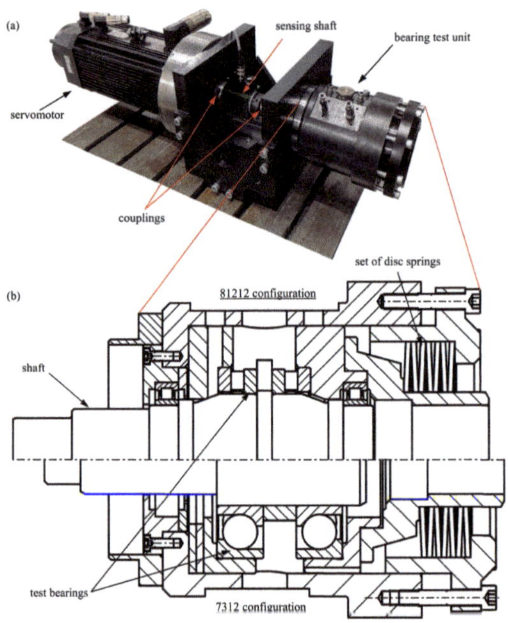

Figure 53: Modified FE8 apparatus from [132], (a) Modified FE8 rig for testing grease-lubricated 81212 CRTB bearings under oscillating conditions; (b) section of modified FE8 test unit for 81212 (upper half) and 7312 (lower half) configuration.

7 Rheological behavior of Lubricating greases

7.1 Introduction

Rheology, as a science describing the flow behavior of liquids and the deformation behavior of solids, is of particular importance for investigating the tribological behavior of viscoelastic materials. This applies to lubricants as well as grease-lubricated friction pairs. Against the backdrop of tribological stress, the time-dependent behavior of the structure-viscous properties of the materials under investigation is central to the studies presented here.

For the representation of the flow properties of different substances, the use of a simple and practical flow model is desirable (see [133]). In the case of simple shear, parallel plane layers slide over each other (with layer thickness assumed to be infinitely small), forming a laminar flow. This simple model is determined by the velocity gradient $\partial u/\partial y$.

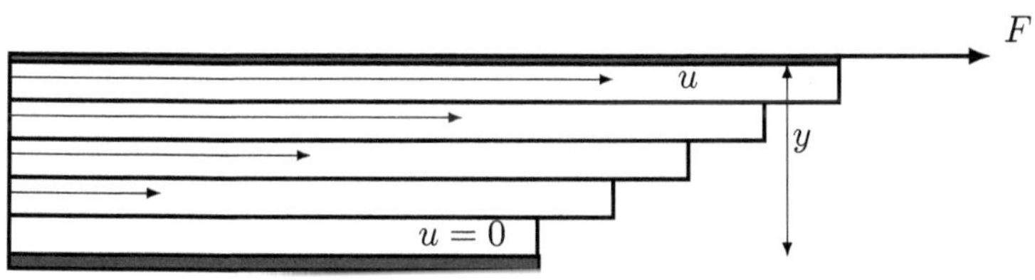

Figure 54: The two-plate-model

Newtonian Fluids For an illustrative representation of the conceptual model characterizing a Newtonian fluid, consider the so-called plate experiment. In this setup, there is a sample substance between a stationary plate and a moving plate (see Figure 54).

Within this substance, a laminar flow is assumed, where the outermost layer slides at the same velocity as the moving plate. The bottom layer remains stationary ($u = 0$) corresponding to the motionless second plate. The change in velocity across the distance between the plates, $\partial u/\partial y$, is referred to as the shear rate $\dot{\gamma}$. If there exists a linear relationship between the shear rate $\dot{\gamma}$ and the shear stress τ (force per unit area), it is termed a *Newtonian fluid.*

$$\tau = \eta \cdot \dot{\gamma} \tag{104}$$

Here, the proportionality factor η represents the dynamic viscosity [mPas] and, for *Newtonian fluids*, it is independent of the shear rate.

Non-Newtonian Fluids In *non-Newtonian fluids*, there is no linear relationship between $\dot{\gamma}$ and τ. Instead, the dynamic viscosity has a series of dependencies: $\eta(\dot{\gamma}(t, t_0))$ [133].

Due to the relevant time-dependent flow behavior in lubricating greases, a rough distinction can be made between *time-independent non-Newtonian behavior* and *time-dependent non-Newtonian behavior*. The former is characterized by a dependency of the viscosity function on the instantaneous value of $\dot{\gamma}$. The latter is distinguished by a dependence of the viscosity function on the shear rate and the duration of stress.

Time-independent Non-Newtonian Flow Behavior This group includes *structure-viscous flow behavior* . If the dispersed phase is given a certain structure, the viscosity decreases with increasing shear stress τ or increasing shear rate $\dot{\gamma}$, and is referred to as structure-viscous [134]. The term *structure-viscous* or structural viscosity was coined by OSTWALD in 1925. The term *pseudoplastic* is also frequently used [134],[133].

When considering the viscosity behavior in relation to the shear rate, two approximately constant viscosity ranges are found for $0 < \dot{\gamma} < \infty$.

$$\eta(\dot{\gamma} \longrightarrow 0) = \eta_0 \tag{105}$$

$$\eta(\dot{\gamma} \longrightarrow \infty) = \eta_\infty < \eta_0 \tag{106}$$

Different terms have developed for these approximately constant ranges and the associated viscosities. The apparent independence from the shear rate led to the terms *lower Newtonian range* and *upper Newtonian range* for very small or very large $\dot{\gamma}$ values (Figure 55). The viscosity at very small values of $\dot{\gamma}$ is referred to as *initial viscosity, zero viscosity, rest viscosity, lower Newtonian limit viscosity.* In contrast, the viscosity at very high shear rates is referred to as *final viscosity, upper Newtonian limit viscosity* (also see Figure 55).

- Plastic flow behavior

Materials exhibiting a yield stress τ_y (origin of curves a, b, c in Figure 56) possess a *plastic flow behavior* . Below this yield stress, predominantly elastic deformations occur, while above this shear stress range, the viscous component predominates. The behavior after exceeding the yield stress can then be represented according to a Newtonian fluid b (*Bingham body*), a dilatant a, or a viscoelastic substance c. For the considerations made here, the *structured viscous* behavior following the attainment of τ_y is relevant.

The yield stress τ_Y is a macroscopic quantity. It is not a material parameter but device-dependent! This means that measured yield stresses can only be compared if the same devices have been used. A new generation of measuring devices and sensors will lead to new values for the yield stress. Therefore, the yield stress describes the sensitivity of the device with which a determination is made. Microscopically, it is assumed that a yield stress does not exist.

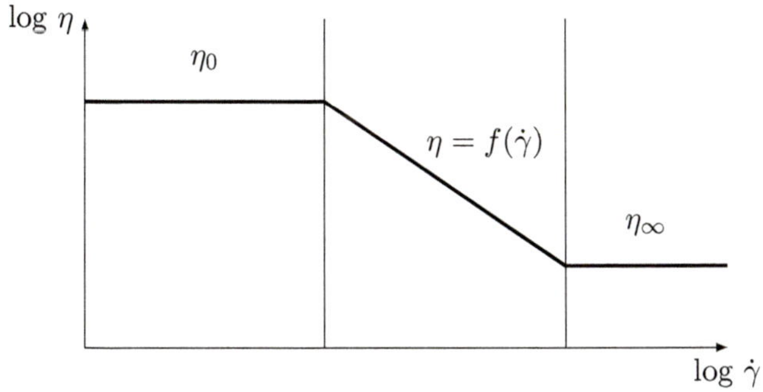

Figure 55: Behavior of a structured liquid

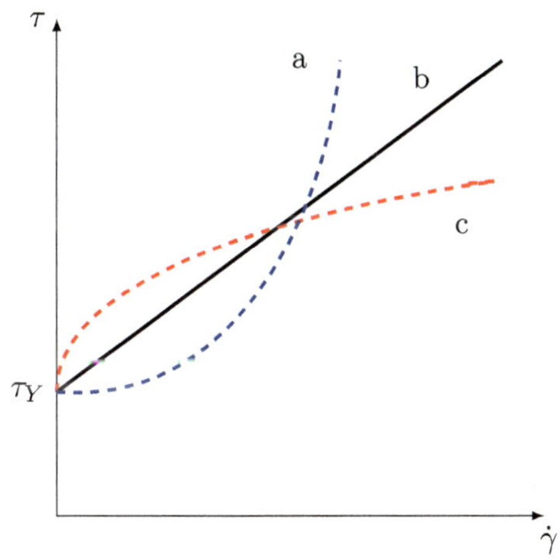

Figure 56: Flow behavior of b) Newtonian fluid, c) structured viscous fluid, a) dilatant fluid. Here in the presence of a yield stress τ_y

On time-dependent Non-Newtonian flow behavior

- Viscoelastic flow behavior

Viscoelastic fluids, as the name suggests, exhibit both viscous and elastic behavior. Examples of viscoelastic behavior of substances can be found in [133]. Here are some examples:

1. A stirrer rotating at a constant speed in a viscoelastic fluid will recoil in the original direction of rotation when the driving torque is suddenly removed.

2. When a stretched liquid filament - such as during the spinning of a plastic fiber - is cut, the two ends spontaneously retract with an increase in diameter.

Interpreting these behaviors highlights the fundamentally different properties of viscoelastic, purely viscous, and purely elastic substances. For a purely elastic material, the time t at which the associated deformation γ is considered is relevant to describe the stress τ. The stress for a purely viscous material is determined by the instantaneous deformation rate $\dot{\gamma}$. To describe the stress state of a viscoelastic fluid, the time evolution (the immediate one) of the deformation must be taken into account.

Viscoelastic fluids exhibit normal stress effects as a special feature. When a non-Newtonian fluid is in a shearing flow, different normal stresses occur in all three directions [135]. Here,

$$\tau_{11} \neq \tau_{22} \neq \tau_{33} \tag{107}$$

In contrast to Newtonian fluids, where [136] states

$$\tau_{11} = \tau_{22} = \tau_{33} = -p \tag{108}$$

To eliminate the influence of hydrostatic pressure when observing normal stress effects (in non-Newtonian fluids), the normal stress differences are formed [137].

$$N_1 = \tau_{11} - \tau_{22} \tag{109}$$

$$N_2 = \tau_{22} - \tau_{33} \tag{110}$$

Here, N_1 and N_2 are characteristic material functions for a non-Newtonian fluid. A well-known example of the occurrence of normal stress effects is the so-called *Weissenberg effect* [138]. Similarly, the *rod climbing* effect is also attributed to the occurrence of normal stresses [133].

Linear viscoelastic flow behavior

A viscoelastic fluid behaves *linearly* when, according to [139],

1. all irreversible components of the flow process behave according to the Newtonian approach (see DIN 1342); this applies to all associated components of stress and deformation velocity.

2. all reversible components of shear deformation behave according to Hooke's law; this applies to all associated components of stress and deformation.

If a parameter of the flow process deviates from the conditions outlined above, the fluid behaves *nonlinearly*.

- Thixotropic flow behavior

The term *thixotropy* was coined by PETERFI [140], who used it in the study of sea urchin eggs. The term is derived from *thixis* meaning touch and *trepo* meaning to turn, alter.

First observed in 1923 by A. SZEGVARI and E. SCHALEK [141], [142], the phenomenon of an isothermal, reversible gel-sol transition was systematically and comprehensively investigated by H. FREUNDLICH [143] .

Thixotropic behavior occurs when the viscosity of a substance decreases under continuous shear stress and constant conditions, rebuilding after a period of rest. In contrast to structural viscosity, a thixotropic substance requires a certain amount of time (which can be considerable under certain circumstances) to rebuild its observable viscosity (see also Figure (57)). This rest time is often much longer than the shear time for many substances [144].

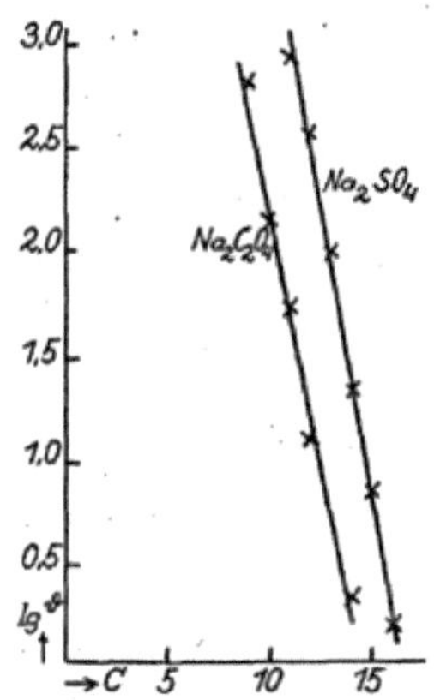

Figure 57: Abscissa: Concentration of added electrolytes, Ordinate: Logarithms of solidification times [143]

A more detailed appreciation of H. FREUNDLICH and a detailed presentation of investigations into thixotropic phenomena are provided in Chapter (9.3.7) in the context of the lubricating grease behavior discussed here.

- Rheopectic Flow Behavior

The term *rheopexy* or *rheopectic* was coined in 1935 by H. FREUNDLICH and F. JULIUSBURGER [145]. They wrote: "We propose to call a thixotropic sol that can solidify through gentle motion rheopectic, and the phenomenon itself rheopexy ". Substances exhibiting rheopectic flow behavior during shearing over time show an increase in viscosity that tends towards a finite value. A subsequent period of rest then leads to a decrease in viscosity again.

7.2 On the Rheology of Greases

For a general characterization of the rheological behavior of greases, the following observations can be made:

- Greases are highly *viscoelastic* lubricants.
- The term *grease thixotropy* indicates that, unlike *thixotropy*, greases exhibit weak structure formation after a shearing stress and a subsequent (considerable) resting period.
- From a macroscopic perspective, greases have a *yield stress* τ_Y.
- Friction in grease decreases with increasing shear rate $\dot{\gamma}$ (for $\tau > \tau_y$), indicating a *structure-viscous* flow behavior.

In the following sections, rheological models will be discussed that are relevant for studying grease behavior. The valuable works of A. DUNKER [146] will be utilized.

7.2.1 Rheological Models for Plastic-Structure-Viscous Flow Behavior

Bingham Model

The BINGHAM-model describes a substance that behaves like a solid at low shear stress and like a liquid at high $\dot{\gamma}$ values [147]. Until the yield stress τ_f is reached, the substance is only elastically deformed [134]. Plastic deformation occurs after exceeding the yield stress, after which Newtonian behavior is observed.

- Shear Stress-Velocity Gradient

The flow law of the Bingham model for stresses greater than the yield stress is given by:

$$\tau(\dot{\gamma}) = \tau_f + \eta_p \cdot \dot{\gamma} \tag{111}$$

Therein,

$\tau(\dot{\gamma})$ = Shear stress at $\dot{\gamma}$ [Pa]; τ_f = Yield stress [Pa]; $\dot{\gamma}$ = Shear rate [s^{-1}]; η_p = Plastic viscosity [Pas]

REINER [134] points out that the signs of the expressions on the right side must be the same.

Limit Value Analysis

The following limit value results:

$$\lim_{\dot{\gamma}\longrightarrow\infty} \tau(\dot{\gamma}) = \lim_{\dot{\gamma}\longrightarrow\infty} (\tau_f + \eta_p \cdot \dot{\gamma}) = \infty \tag{112}$$

The BINGHAM-model approaches infinity as $\dot{\gamma} \longrightarrow \infty$. However, studies (e.g., [148]) suggest the existence of a limiting shear stress τ_L that cannot be exceeded. For $\dot{\gamma} = 0$, the equation is defined and yields

$$\tau(\dot{\gamma} = 0) = \tau_f \tag{113}$$

The equation indicates the yield stress τ_f for $\dot{\gamma} = 0$, implying that plastic flow occurs only after exceeding the critical value.

Slope

The slope of the BINGHAM-model is given by:

$$\frac{d\tau}{d\dot{\gamma}} = \eta_p = const. \tag{114}$$

- Apparent viscosity-shear rate

The combination of the BINGHAM approach with the Newtonian model yields

$$\eta'(\dot{\gamma}) = \frac{\tau_f}{\dot{\gamma}} + \eta_p \tag{115}$$

Limit value analysis

The limit value analysis yields:

$$\lim_{\dot{\gamma} \longrightarrow 0} \eta'(\dot{\gamma}) = \lim_{\dot{\gamma} \longrightarrow 0} \left(\frac{\tau_f}{\dot{\gamma}} + \eta_p \right) = \infty \tag{116}$$

It is evident that the apparent viscosity $\eta'(\dot{\gamma})$ becomes infinitely large as the shear rate approaches zero. This is consistent with the fact that lubricating greases also exhibit solid-like properties below the yield stress. For a shear rate approaching infinity, one obtains

$$\lim_{\dot{\gamma} \longrightarrow \infty} \eta\prime(\dot{\gamma}) = \lim_{\dot{\gamma} \longrightarrow \infty} \left(\frac{\tau_f}{\dot{\gamma}} + \eta_p \right) = \eta_p \tag{117}$$

This means that the apparent viscosity approaches the plastic viscosity η_p. [147] and [149] refer to this state as η_∞ or η_{Rest}.

Slope

The slope for the apparent viscosity is obtained by taking the derivative with respect to $\dot{\gamma}$, resulting in:

$$\frac{d\eta\prime}{d\dot{\gamma}} = -\frac{\tau_f}{\dot{\gamma}^2} \tag{118}$$

Then, the following *limit value* arises:

$$\lim_{\dot{\gamma} \longrightarrow \infty} \frac{d\eta'}{d\dot{\gamma}} = \lim_{\dot{\gamma} \longrightarrow \infty} \left(-\frac{\tau_f}{\dot{\gamma}^2} \right) = 0 \tag{119}$$

Equation by Casson

Unlike the BINGHAM approach, the equation by CASSON describes a non-linear dependence of the shear stress on the shear rate above the yield point.

In contrast to the BINGHAM approach, the equation by CASSON describes a non-linear dependence of the shear stress on the shear rate above the yield point [137]. It is often used as an approximation for the flow behavior of lubricating greases. The general expression is:

$$\tau^n - \tau_f^n = (\eta_C \cdot \dot{\gamma})^n \tag{120}$$

or

$$\tau(\dot{\gamma}) = \left(\tau_f^n + (\eta_C \cdot \dot{\gamma})^n \right)^{\frac{1}{n}} \tag{121}$$

In which:

$\tau(\dot{\gamma})$= Shear stress at $\dot{\gamma}$ [Pa]; τ_f= Yield stress [Pa]; $\dot{\gamma}$= Shear rate [s^{-1}]; η_C= Viscosity coefficient [Pas]; n= rheological parameter [-]

The *limit value analysis* yields the following results:

$$\lim_{\dot{\gamma} \longrightarrow \infty} \tau(\dot{\gamma}) = \lim_{\dot{\gamma} \longrightarrow \infty} \left(\left[\tau_f^n + (\eta_C \cdot \dot{\gamma})^n \right]^{\frac{1}{n}} \right) = \infty \tag{122}$$

For infinitely large shear rates, the shear stress also tends to infinity. For values of $\dot{\gamma} = 0$, we obtain

$$\tau(\dot{\gamma} = 0) = \tau_f \tag{123}$$

This means that no flow occurs before reaching the yield stress.

Slope

$$\frac{d\tau}{d\dot{\gamma}} = \left[\left(\frac{\tau_f}{\dot{\gamma}}\right)^n + \eta_C^n\right]^{\frac{1-n}{n}} \cdot \eta_C^n \tag{124}$$

Thus, the following *limit value* is obtained:

$$\lim_{\dot{\gamma}\longrightarrow\infty} \frac{d\tau}{d\dot{\gamma}} = \lim_{\dot{\gamma}\longrightarrow\infty} \left(\left[\left(\frac{\tau_f}{\dot{\gamma}}\right)^n + \eta_C^n\right]^{\frac{1-n}{n}} \cdot \eta_C^n\right) = \eta_C \tag{125}$$

This means that the shear stress equation $\tau(\dot{\gamma})$ exhibits Newtonian behavior with constant slope for the considered limit case.

- Apparent viscosity-shear rate

The apparent viscosity is obtained by combining CASSON ($\tau(\dot{\gamma})$) and NEWTON as follows:

$$\eta\prime(\dot{\gamma}) = \left[\left(\frac{\tau_f}{\dot{\gamma}}\right)^n + \eta_C^n\right]^{\frac{1}{n}} \tag{126}$$

where $n < 1$.
A *limit value analysis* yields:

$$\lim_{\dot{\gamma}\longrightarrow\infty} \eta\prime(\dot{\gamma}) = \lim_{\dot{\gamma}\longrightarrow\infty} \left(\left[\left(\frac{\tau_f}{\dot{\gamma}}\right)^n + \eta_c^n\right]^{\frac{1}{n}}\right) = \infty \tag{127}$$

This implies that the apparent viscosity becomes infinitely large for $\dot{\gamma} = 0$. This models the behavior below the yield stress (solid-like properties).

$$\lim_{\dot{\gamma}\longrightarrow\infty} \eta\prime(\dot{\gamma}) = \lim_{\dot{\gamma}\longrightarrow\infty} \left(\left[\left(\frac{\tau_f}{\dot{\gamma}}\right)^n + \eta_C^n\right]^{\frac{1}{n}}\right) = \eta_C \tag{128}$$

For infinitely large shear rates, η_C (CASSON viscosity) is obtained, indicating Newtonian behavior. In [147], CZARNY equates η_C with the viscosity η_∞.

Slope

The slope of the apparent viscosity curve provides the first derivative with respect to $\dot{\gamma}$.

$$\frac{d\eta\prime}{d\dot{\gamma}} = \left[\left(\frac{\tau_f}{\dot{\gamma}}\right)^n + \eta_C^n\right]^{\frac{1}{n}-1} \cdot \left(-\frac{\tau_f^n}{\dot{\gamma}^{n+1}}\right) \tag{129}$$

Then, the slope yields the following *limit value*

$$\lim_{\dot{\gamma}\longrightarrow\infty} \frac{d\eta\prime}{d\dot{\gamma}} = \lim_{\dot{\gamma}\longrightarrow\infty} \left(\left[\left(\frac{\tau_f}{\dot{\gamma}}^n\right) + \eta_C^n\right]^{\frac{1}{n}-1} \cdot \left(-\frac{\tau_f^n}{\dot{\gamma}^{n+1}}\right)\right) = 0 \tag{130}$$

For the limiting case $\dot{\gamma} \to \infty$, the apparent viscosity exhibits Newtonian behavior with a slope of 0.

Bauer-Equation by Åström/Höglund

The temperature dependence of the parameters a and b in the BAUER-equation [150]:

$$\tau(\dot{\gamma}) = \tau_f + a \cdot \dot{\gamma} + b \cdot \dot{\gamma}^n \tag{131}$$

is formulated by ÅSTRÖM and HÖGLUND in an extended approach [148]:

$$\tau(\dot{\gamma}, T) = \tau_f + a_T \cdot \dot{\gamma} + b \cdot e^{(c(T-T_0))} \cdot \dot{\gamma}^n \tag{132}$$

Therein,
$\tau(\dot{\gamma})$= Shear stress at $\dot{\gamma}$[Pa]; τ_f= Yield stress [Pa]; $\dot{\gamma}$= Shear rate [s^{-1}]; a_T= Base oil viscosity at ϑ[Pas]; b= Parameter [Pas^n]; c= Temperature coefficient [K^{-1}]; T= Temperature [K]; T_0= Reference temperature [K]

The BAUER-equation is extended by the temperature coefficient c. Here, the parameter b is also subjected to temperature dependence through the expression $(T - T_0)$.

Limit Value Analysis

For the limit value analysis, we have:

$$\lim_{\dot{\gamma}\longrightarrow\infty} \tau(\dot{\gamma}, T) = \lim_{\dot{\gamma}\longrightarrow\infty} (\tau_f + a_T \cdot \dot{\gamma} + b \cdot e^{(c(T-T_0))} \cdot \dot{\gamma}^n) = \infty \tag{133}$$

$$\tau(\dot{\gamma} = 0) = \tau_f \tag{134}$$

For both limit cases, the original BAUER-equation is thus satisfied.
Slope
The examination of the slope yields:

$$\frac{d\tau}{d\dot{\gamma}} = a_T + n \cdot b \cdot e^{(c(T-T_0))} \cdot \dot{\gamma}^{n-1} \tag{135}$$

with $n < 1$.
Therefore, the slope has the following *limit value*:

$$\lim_{\dot{\gamma} \longrightarrow \infty} \frac{d\tau}{d\dot{\gamma}} = \lim_{\dot{\gamma} \longrightarrow \infty} (a_T + n \cdot b \cdot e^{(c(T-T_0))} \cdot \dot{\gamma}^{n-1}) = a_T \tag{136}$$

- Apparent viscosity-shear rate

With the ÅSTRÖM/HÖGLUND approach and the Newtonian relationship, the apparent viscosity is given by:

$$\eta\prime(\dot{\gamma}, T) = \frac{\tau_f}{\dot{\gamma}} + a_T + b \cdot e^{(c(T-T_0))} \cdot \dot{\gamma}^{n-1} \tag{137}$$

where $n < 1$.
Limit value analysis
For the apparent viscosity:

$$\lim_{\dot{\gamma} \longrightarrow 0} \eta\prime(\dot{\gamma}, T) = \lim_{\dot{\gamma} \longrightarrow 0} \left(\frac{\tau_f}{\dot{\gamma}} + a_t + b \cdot e^{(c(T-T_0))} \cdot \dot{\gamma}^{n-1} \right) = \infty \tag{138}$$

$$\lim_{\dot{\gamma} \longrightarrow \infty} \eta\prime(\dot{\gamma}, T) = \lim_{\dot{\gamma} \longrightarrow \infty} \left(\frac{\tau_f}{\dot{\gamma}} + a_t + b \cdot e^{(c(T-T_0))} \cdot \dot{\gamma}^{n-1} \right) = a_T \tag{139}$$

Slope

$$\frac{d\eta\prime}{d\dot{\gamma}} = -\frac{\tau_f}{\dot{\gamma}^2} + (n-1) \cdot b \cdot e^{(c(T-T_0))} \cdot \dot{\gamma}^{n-2} \tag{140}$$

for $n < 1$.
For the slope, the following *limit value* is obtained:

$$\lim_{\dot{\gamma} \longrightarrow \infty} \frac{d\eta\prime}{d\dot{\gamma}} = \lim_{\dot{\gamma} \longrightarrow \infty} \left(-\frac{\tau_f}{\dot{\gamma}^2} + (n-1) \cdot b \cdot e^{(c(T-T_0))} \cdot \dot{\gamma}^{n-2} \right) = 0 \tag{141}$$

An application example is presented in [148].

Modified Bingham Model by Bair [151]

- Shear Stress - Shear Rate - Ratio

To investigate the behavior of greases under high pressures, BAIR modifies the BINGHAM-model.

$$\tau(\dot{\gamma})^q = \tau_f^q + (g \cdot \eta_G \cdot \dot{\gamma})^q \tag{142}$$

with $q < 1$.

We can further write

$$\tau(\dot{\gamma}) = \left[\tau_f^q + (g \cdot \eta_G \cdot \dot{\gamma})^q\right]^{\frac{1}{q}} \tag{143}$$

Therein:

$\tau(\dot{\gamma})$= Shear stress at $\dot{\gamma}$ [Pa]; τ_f= Yield stress [Pa]; $\dot{\gamma}$= Shear rate [s^{-1}]; η_G= Base oil viscosity [Pas]; g= Parameter describing viscosity reduction [$-$]; q= Dimensionless rheological parameter [-]

BAIR characterizes the approach (143) as particularly suitable for describing grease behavior at high pressures. The distinction from the CASSON-equation lies solely in the introduction of the factor g, so that the exponent q is to be treated as in equation (121].

The influence of pressure and temperature on the base oil viscosity is described by [152] according to [151]

$$\eta_G(T,p) = \eta_g \cdot exp\left(\frac{-2.3 \cdot C_1(T - T_g) \cdot F}{C_2 + (T - T_g) \cdot F}\right) \tag{144}$$

where

$$T_g = T_{g0} + A_1 \cdot ln(1 + A_2 \cdot p) \tag{145}$$

and

$$F = 1 - B_1 \cdot (1 + B_2 \cdot p) \tag{146}$$

Here,

η_G = Base oil viscosity [Pas]; T= Temperature [K], T_g= Glass transition temperature [K]; T_{g0}= Glass transition temperature at p = 0 [K]; p= Pressure [Pa]; F= Relative free volume expansion [-]; A_1= Constant [K]; A_2 Constant [Pa^{-1}]; B_1= Constant [-]; B_2= Constant [Pa^{-1}]; C_1= Constant [-]; C_2= Constant [K]

The determination of material parameters (constants A to C) is carried out for each base oil with experimentally determined flow curves and an adjustment of Equation (144).

Limit value analysis

For Equation (143), the following consideration is made

$$\lim_{\dot{\gamma}\longrightarrow\infty} \tau(\dot{\gamma}) = \lim_{\dot{\gamma}\longrightarrow\infty} \left[\left(\tau_f^q + (g \cdot \eta_g \cdot \dot{\gamma})^q\right]^{\frac{1}{q}}\right] = \infty \tag{147}$$

It turns out that Equation (143) also causes the shear stress to approach infinity for an infinitely large shear rate.

For a shear rate $\dot{\gamma} = 0$, Equation (143) yields the shear stress τ_f. This means that flow occurs only after exceeding the yield point.

$$\tau\left(\dot{\gamma} = 0\right) = \tau_f \tag{148}$$

Slope

For the slope of the BAIR-equation, we obtain

$$\frac{d\tau}{d\dot{\gamma}} = \left[\left(\frac{\tau_f}{\dot{\gamma}}\right)^q + (g \cdot \eta_G)^q\right]^{\frac{1-q}{q}} \cdot (g \cdot \eta_G)^q \tag{149}$$

and further, for Equation (149), we get

$$\lim_{\dot{\gamma}\longrightarrow\infty} \frac{d\tau}{d\dot{\gamma}} = \lim_{\dot{\gamma}\longrightarrow\infty} \left\{\left[\left(\frac{\tau_f}{\dot{\gamma}}\right)^q + (g \cdot \eta_G)^q\right]^{\frac{1-q}{q}} \cdot (g \cdot \eta_G)^q\right\} = g \cdot \eta_G \tag{150}$$

It is shown that for $\dot{\gamma} \longrightarrow \infty$ a Newtonian flow behavior with a constant slope is observed.

- Apparent Viscosity - Shear Rate Relationship

For the apparent viscosity $\eta\prime\left(\dot{\gamma}\right)$, Equation (143) is used in the Newtonian approach.

$$\eta\prime(\dot{\gamma}) = \left[\left(\frac{\tau_f}{(\dot{\gamma})}\right)^q + (g \cdot \eta_G)^q\right]^{\frac{1}{q}} \tag{151}$$

Limit Value Analysis

The limit value analysis yields the following results.

$$\lim_{\dot{\gamma}\longrightarrow 0} \eta\prime(\dot{\gamma}) = \lim_{\dot{\gamma}\longrightarrow 0} \left\{ \left[\left(\frac{\tau_f}{(\dot{\gamma})} \right)^q + (g \cdot \eta_G)^q \right]^{\frac{1}{q}} \right\} = \infty \tag{152}$$

Below the yield stress, the solid properties dominate, and for $\dot{\gamma} \longrightarrow 0$, the apparent viscosity is interpreted as infinitely large. Furthermore,

$$\lim_{\dot{\gamma}\longrightarrow\infty} \eta\prime(\dot{\gamma}) = \lim_{\dot{\gamma}\longrightarrow\infty} \left\{ \left[\left(\frac{\tau_f}{(\dot{\gamma})} \right)^q + (g \cdot \eta_G)^q \right]^{\frac{1}{q}} \right\} = g \cdot \eta_G \tag{153}$$

This means that the apparent viscosity approaches the limit $g \cdot \eta_G$ for $\dot{\gamma} \longrightarrow \infty$ and Equation (151) shows a Newtonian behavior in this case. It is excluded that the viscosity of the lubricant develops to a level below the base oil viscosity. Thus, $g > 1$.

Slope

The slope according to Equation (151) is given by

$$\frac{d\eta\prime}{d\dot{\gamma}} = \left[\left(\frac{\tau_f}{\dot{\gamma}} \right)^q + (g \cdot \eta_G)^q \right]^{\frac{1}{q}-1} \cdot \left(-\frac{\tau_f^q}{\dot{\gamma}^{q+1}} \right) \tag{154}$$

The investigation of Equation (154) yields the following *limit value*:

$$\lim_{\dot{\gamma}\longrightarrow\infty} \frac{d\eta\prime}{d\dot{\gamma}} = \lim_{\dot{\gamma}\longrightarrow\infty} \left[\left(\frac{\tau_f}{\dot{\gamma}} \right)^q + (g \cdot \eta_G)^q \right]^{\frac{1}{q}-1} \cdot \left(-\frac{\tau_f^q}{\dot{\gamma}^{q+1}} \right) = 0 \tag{155}$$

The apparent viscosity modeled by the relation (151) shows a Newtonian behavior for this limit value.

Herschel-Bulkley Equation

- Shear stress - shear rate relationship

The combination of the BINGHAM-model with the OSTWALD-DE WAELE power law leads to the HERSCHEL-BULKLEY-equation [153]

$$\tau - \tau_f = \kappa \cdot \dot{\gamma}^n \tag{156}$$

and thus

$$\tau(\dot{\gamma}) = \tau_f + \kappa \cdot \dot{\gamma}^n \tag{157}$$

Where:
$\tau(\dot{\gamma})$ = shear stress at $\dot{\gamma}$ [Pa]; τ_f = yield stress [Pa]; $\dot{\gamma}$ = shear rate [s^{-1}]; κ = consistency index [Pa·s^n]; n = dimensionless rheological parameter [-]

The formulation of Equation (157) requires that the terms on the right-hand side have the same sign. A dimensional concern arises since the dimension of the parameter κ depends on the value of the exponent n.

Limit Value Analysis

With Equation (157) we get

$$\lim_{\dot{\gamma}\longrightarrow\infty} \tau(\dot{\gamma}) = \lim_{\dot{\gamma}\longrightarrow\infty} (\tau_f + \kappa \cdot \dot{\gamma}^n) = \infty \tag{158}$$

Considering $\dot{\gamma} = 0$ we get

$$\tau(\dot{\gamma} = 0) = \tau_f \tag{159}$$

This means that flow starts only after reaching the yield stress.

Slope

For the slope we get

$$\frac{d\tau}{d\dot{\gamma}} = n \cdot \kappa \cdot \dot{\gamma}^{n-1} \tag{160}$$

where $n < 1$. This gives the following *limit value*:

$$\lim_{\dot{\gamma}\longrightarrow\infty} \frac{d\tau}{d\dot{\gamma}} = \lim_{\dot{\gamma}\longrightarrow\infty} \left(n \cdot \kappa \cdot \dot{\gamma}^{n-1}\right) = 0 \tag{161}$$

This means that for this limit case the viscosity "disappears". This is a circumstance that appears problematic.

- Apparent Viscosity - Shear Rate Relationship

The description of the apparent viscosity is given by

$$\eta'(\dot{\gamma}) = \frac{\tau_f}{\dot{\gamma}} + \kappa \cdot \dot{\gamma}^{n-1} \tag{162}$$

with $n < 1$.

Limit Value Analysis

The following limit values are obtained:

$$\lim_{\dot{\gamma}\to 0} \eta'(\dot{\gamma}) = \lim_{\dot{\gamma}\to 0} \left(\frac{\tau_f}{\dot{\gamma}} + \kappa \cdot \dot{\gamma}^{n-1}\right) = \infty \tag{163}$$

This limit value analysis accounts for the notion that lubricating greases exhibit solid-like properties below the yield stress.

$$\lim_{\dot{\gamma}\to \infty} \eta'(\dot{\gamma}) = \lim_{\dot{\gamma}\to \infty} \left(\frac{\tau_f}{\dot{\gamma}} + \kappa \cdot \dot{\gamma}^{n-1}\right) = 0 \tag{164}$$

It is evident that with Equation (162), the apparent viscosity becomes 0 as $\dot{\gamma}$ approaches 0. This would imply a decrease in grease viscosity below the base oil viscosity and poses a limitation on the applicability of the HERSCHEL-BULKLEY-equation.

Slope

Next, the slope of the viscosity curve described by Equation (162) is presented as follows:

$$\frac{d\eta'}{d\dot{\gamma}} = -\frac{\tau(\dot{\gamma})}{\dot{\gamma}^2} + \kappa \cdot (n-1) \cdot \dot{\gamma}^{n-2} \tag{165}$$

and the subsequent *limit value* is obtained as:

$$\lim_{\dot{\gamma}\longrightarrow \infty} \frac{d\eta'}{d\dot{\gamma}} = \lim_{\dot{\gamma}\longrightarrow \infty} \left(-\frac{\tau_f}{\dot{\gamma}^2} + \kappa \cdot (n-1) \cdot \dot{\gamma}^{n-2}\right) = 0 \tag{166}$$

Sisko Equation

- Shear Stress - Shear Rate - Relationship

$$\tau(\dot{\gamma}) = \eta_\infty \cdot \dot{\gamma} + \kappa \cdot \dot{\gamma}^n \tag{167}$$

For $n < 1$.

Where

$\tau(\dot{\gamma})$ = Shear stress at $\dot{\gamma}$ [Pa]; $\dot{\gamma}$ = Shear rate [s^{-1}]; η_∞ = Structural grease viscosity at high shear rate [$Pa \cdot s$]; κ = Parameter [$Pa \cdot s^n$]; n = Dimensionless rheological parameter [-]

The signs of the terms on the right-hand side of Equation (167) must be the same. The dimension of the parameter κ changes with the value of the exponent n. There is also a dimensional inconsistency here.

Limit Value Analysis

For Equation (167), we find

$$\lim_{\dot{\gamma} \longrightarrow \infty} \tau(\dot{\gamma}) = \lim_{\dot{\gamma} \longrightarrow \infty} (\eta_\infty \cdot \dot{\gamma} + \kappa \cdot \dot{\gamma}^n) = \infty \tag{168}$$

The SISKO equation exhibits a shear stress that also tends towards infinity as $\dot{\gamma} \longrightarrow \infty$, and for $\dot{\gamma} \longrightarrow 0$, Equation (167) yields

$$\tau(\dot{\gamma} = 0) = 0 \tag{169}$$

Here, a lower limit of application of the treated relationship is shown. In [154], it is mentioned that the SISKO equation can be applied for ranges of $\dot{\gamma} = 10^{-2}s^{-1}$ to approximately $10^4 s^{-1}$.

Slope

For the slope, we can write

$$\frac{d\tau}{d\dot{\gamma}} = \eta_\infty + n \cdot \kappa \cdot \dot{\gamma}^{n-1} \tag{170}$$

where $n < 1$.

The *limit value analysis* yields:

$$\lim_{\dot{\gamma}\longrightarrow\infty} \frac{d\tau}{d\dot{\gamma}} = \lim_{\dot{\gamma}\longrightarrow\infty} \left(\eta_{\infty} + n \cdot \kappa \cdot \dot{\gamma}^{n-1}\right) = \eta_{\infty} \tag{171}$$

As $\dot{\gamma} \longrightarrow \infty$, the flow curve of Equation (167) exhibits *Newtonian* behavior.

- Apparent Viscosity - Shear Rate - Ratio

In connection with Equation (104), the apparent viscosity is given by

$$\eta\prime(\dot{\gamma}) = \eta_{\infty} \cdot \dot{\gamma} + \kappa \cdot \dot{\gamma}^{n-1} \tag{172}$$

with $n < 1$.

Limit values

The following limit values are described:

$$\lim_{\dot{\gamma}\longrightarrow 0} \eta\prime(\dot{\gamma}) = \lim_{\dot{\gamma}\longrightarrow 0} \left(\eta_{\infty} + n \cdot \kappa \cdot \dot{\gamma}^{n-1}\right) = \infty \tag{173}$$

Here, too, the concept that lubricating greases exhibit solid-like properties below a yield threshold is reflected.

$$\lim_{\dot{\gamma}\longrightarrow\infty} \eta\prime(\dot{\gamma}) = \lim_{\dot{\gamma}\longrightarrow\infty} \left(\eta_{\infty} + n \cdot \kappa \cdot \dot{\gamma}^{n-1}\right) = \eta_{\infty} \tag{174}$$

The limit analysis for this case yields a Newtonian behavior. In [154], [148], [155], it is noted that for η_{∞}, the base oil viscosity can be used as a good approximation.

Slope

For the slope of the apparent viscosity curve described by Equation (172), we have

$$\frac{d\eta\prime}{d\dot{\gamma}} = \kappa \cdot (n-1) \cdot \dot{\gamma}^{n-2} \tag{175}$$

Thus, the following *limit value* is obtained:

$$\lim_{\dot{\gamma}\longrightarrow\infty} \frac{d\eta\prime}{d\dot{\gamma}} = \lim_{\dot{\gamma}\longrightarrow\infty} \left(\kappa \cdot (n-1) \cdot \dot{\gamma}^{n-2}\right) = 0 \tag{176}$$

We observe a Newtonian behavior with a slope of 0.

Stanulov et al. Equation

- Shear Stress - Shear Rate - Ratio

STANULOV ET AL. reported on the experimental investigation of lubricating greases with polymer additives and developed the following approach to describe the shear stress behavior.

$$\tau\left(\dot{\gamma}\right) = \tau_f + a \cdot \left(1 - e^{-b\dot{\gamma}}\right) + \eta_\infty \cdot \dot{\gamma} \tag{177}$$

Where:

$\tau(\dot{\gamma})$= Shear stress at $\dot{\gamma}$ [Pa]; τ_f= Yield stress [Pa]; η_∞= Minimum plastic viscosity $[Pa \cdot s]$ a= Necessary stress for the transition of the lubricating grease to the destroyed state $[Pa]$; b= Rate of structure destruction in the transition region $[s]$

According to [156], Equation (177) represents the quantitative change in structural strength with the quantities $(\tau_f + a)$. The constant b describes the rate of structure degradation.

Limit value

Considering Equation (177) yields the following *limit value*

$$\lim_{\dot{\gamma}\longrightarrow\infty} \tau\left(\dot{\gamma}\right) = \lim_{\dot{\gamma}\longrightarrow\infty} \left[\tau_f + a \cdot \left(1 - e^{-b\dot{\gamma}}\right) + \eta_\infty \cdot \dot{\gamma}\right] = \infty \tag{178}$$

That is, the shear stress tends to infinity as $\dot{\gamma} \longrightarrow \infty$. The equation is also defined for $\dot{\gamma} \longrightarrow 0$, yielding

$$\tau\left(\dot{\gamma} = 0\right) = \tau_f \tag{179}$$

The interpretation suggests flow initiation only after reaching a yield stress τ_f.

Slope

From Equation (177), we find for the investigation of the gradient

$$\frac{d\tau}{d\dot{\gamma}} = a \cdot b \cdot e^{-b \cdot \dot{\gamma}} + \eta_\infty \tag{180}$$

and for the *limit analysis*, we get:

$$\lim_{\dot{\gamma} \longrightarrow \infty} \frac{d\tau}{d\dot{\gamma}} = \lim_{\dot{\gamma} \longrightarrow \infty} \left(a \cdot b \cdot e^{-b \cdot \dot{\gamma}} + \eta_\infty \right) = \eta_\infty \tag{181}$$

which implies a Newtonian behavior of the flow curve according to Equation (177) for the limit case $\dot{\gamma} \longrightarrow \infty$.

- Apparent Viscosity - Shear Rate - Ratio

For the apparent viscosity, we have

$$\eta\prime\left(\dot{\gamma}\right) = \frac{\tau_f}{\dot{\gamma}} + \frac{a}{\dot{\gamma}} \cdot \left(1 - e^{-b \cdot \dot{\gamma}}\right) + \eta_\infty \tag{182}$$

Limit analysis

and for the limit analysis, we get

$$\lim_{\dot{\gamma} \longrightarrow 0} \eta\prime\left(\dot{\gamma}\right) = \lim_{\dot{\gamma} \longrightarrow 0} \left[\frac{\tau_f}{\dot{\gamma}} + \frac{a}{\dot{\gamma}} \cdot \left(1 - e^{-b \cdot \dot{\gamma}}\right) + \eta_\infty\right] = \infty \tag{183}$$

The assumption that lubricating greases exhibit solid-like properties below a flow threshold finds expression here.

$$\lim_{\dot{\gamma} \longrightarrow \infty} \eta\prime\left(\dot{\gamma}\right) = \lim_{\dot{\gamma} \longrightarrow \infty} \left[\frac{\tau_f}{\dot{\gamma}} + \frac{a}{\dot{\gamma}} \cdot \left(1 - e^{-b \cdot \dot{\gamma}}\right) + \eta_\infty\right] = \eta_\infty \tag{184}$$

From Stanulov et al., the parameter η_∞ is referred to as the plastic viscosity. For the considered limit, the relation (182) exhibits a Newtonian behavior.

Slope

We examine the slope of the viscosity curve (Equation (182)).

$$\frac{d\eta\prime}{d\dot{\gamma}} = \frac{1}{\dot{\gamma}^2}\left[-\tau_f - a + a \cdot e^{-b \cdot \dot{\gamma}} \left(b \cdot \dot{\gamma} + 1\right)\right] \tag{185}$$

and the subsequent *limit* is as follows:

$$\lim_{\dot{\gamma}\longrightarrow\infty}\frac{d\eta\prime}{d\dot{\gamma}}=\lim_{\dot{\gamma}\longrightarrow\infty}\left\{\frac{1}{\dot{\gamma}^2}\left[-\tau_f-a+a\cdot e^{-b\cdot\dot{\gamma}}\left(b\cdot\dot{\gamma}+1\right)\right]\right\}=0 \tag{186}$$

which demonstrates a Newtonian behavior (slope = 0) of the modeled viscosity curve.

7.2.2 Models of Time-Dependent Flow Behavior

Bauer Equation

Also proposed by [150] is a suggestion for describing time-dependent flow behavior. (The basis of this approach is a proposal by [157]).

$$\tau(t)=\tau_t+c\cdot t^m \tag{187}$$

with $m<0$.
Where:
$\tau(t)$= Shear stress at time t [Pa]; τ_t= Asymptotic stress value [Pa]; t= Time of exposure [s]; c= Coefficient [Pas^{-m}]; m= Dimensionless exponent

For the parameter referred to here as the *asymptotic stress value*, a variety of designations can be found in the literature. [158] use the term *equilibrium viscosity* for the viscosity state occurring at τ_t, [66] uses the term *equilibrium shear stress*, and [149] introduces the terms *residual shear stress* or *residual viscosity*.

Limit value analysis

$$\lim_{t\longrightarrow 0}\tau(t)=\lim_{t\longrightarrow 0}\left(\tau_t+c\cdot t^m\right)=\infty \tag{188}$$

This means that the shear stress tends to an infinite value for $t=0$ (with finite initial shear stress), which contradicts real conditions.

$$\lim_{t\longrightarrow\infty}\tau(t)=\lim_{t\longrightarrow\infty}\left(\tau_t+c\cdot t^m\right)=\tau_t \tag{189}$$

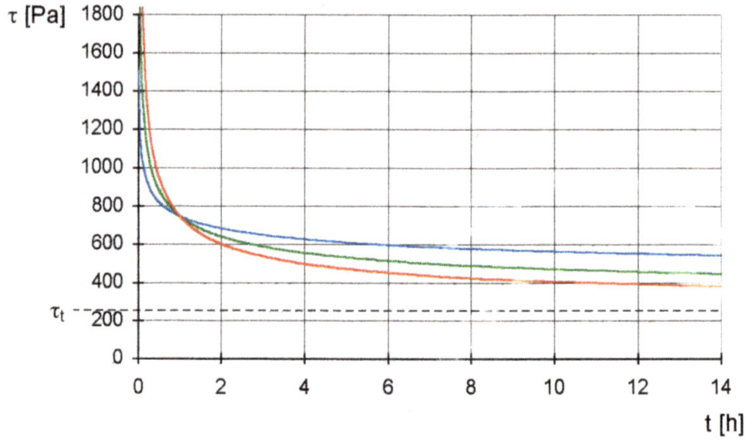

Figure 58: Behavior of Eq.(187) for different values of m (from bottom to top -0.5, -0.35, -0.2 at $t = 14h$). Assumed values are $\tau_t = 250Pa$, $c = 500Pa \cdot h^{-m}$,

As $t \longrightarrow \infty$, the shear stress tends to the asymptotic stress value τ_t. There are formal objections because of the behavior of $t \longrightarrow 0$. The equation is not defined for $t = 0$. The dimensional issue, as generally raised for the OSTWALD-DE WAELSCHE relationship, is also present here since the equation contains a power term.

Slope

With

$$\frac{d\tau}{dt} = m \cdot c \cdot t^{m-1} \tag{190}$$

the following *limit value* is obtained for the slope:

$$\lim_{t \longrightarrow \infty} \frac{d\tau}{dt} = \lim_{t \longrightarrow \infty} \left(m \cdot c \cdot t^{m-1} \right) = 0 \tag{191}$$

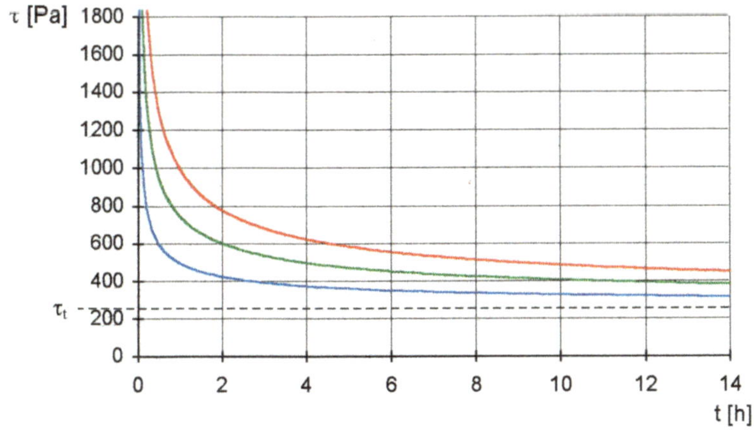

Figure 59: Behavior of Eq.(187) for different values of c ($250 Pa \cdot h^{-m}$, $500 Pa \cdot h^{-m}$, $750 Pa \cdot h^{-m}$). Assumed values are $\tau_t = 250 Pa$, $m = -0.5$,

Czarny Equation

CZARNY developed an equation for constant shear rate $\dot{\gamma}$ and constant temperature ϑ [66]:

$$\tau(t) = \left(k_1 \cdot (m-1) \cdot t \cdot \tau_0^{1-m} + (\tau_0 - \tau_r)^{1-m}\right)^{\frac{1}{1-m}} + \tau_r \tag{192}$$

Where:

$\tau(t)$= Shear stress at time t [Pa]; τ_0= Initial shear stress at $t = 0$ [Pa]; τ_r= Equilibrium shear stress [Pa]; k_1= Coefficient indicating the intensity of structural decay [s^{-1}]; m= Dimensionless exponent characterizing the decay curve; t= Time of exposure [s]

CZARNY uses the term *equilibrium shear stress*, which corresponds to the terminology mentioned earlier with *asymptotic stress* and *residual shear stress.*

Limit Analysis

The following limit is determined:

$$\lim_{t \longrightarrow \infty} \tau(t) = \lim_{t \longrightarrow \infty} \left(\left[k_1(m-1) \cdot t \cdot \tau_0^{1-m} + (\tau_0 - \tau_r)^{1-m}\right]^{\frac{1}{1-m}} + \tau_r\right) = \tau_r \tag{193}$$

This means that the shear stress $\tau(t)$ approaches the equilibrium shear stress τ_r as $t \longrightarrow \infty$.

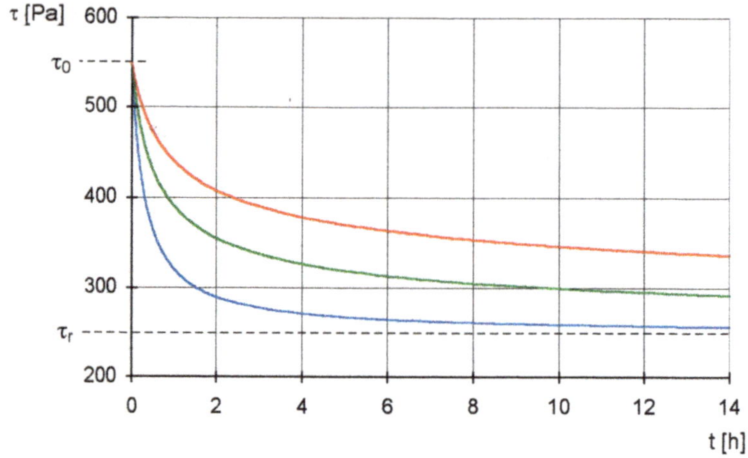

Figure 60: Behavior of Eq.(192) for different values of m (2, 3, 4). Assumed values are $\tau_0 = 550Pa$, $\tau_r = 250Pa$, $k_1 = 6h^{-1}$,

The function is also defined for $t = 0$, yielding the initial shear stress τ_0.

The aforementioned dimensional objection does not apply to the Czarny relationship. However, the dimensionlessness of the parameter k_1 specified by [66] leads to a contradiction (different dimensions of the listed summands). In [146], A. DUNKER suggests assigning this parameter the inverse dimension of the exposure time.

Slope

With

$$\frac{d\tau}{dt} = \frac{1}{1-m}\left(k_1 \cdot (m-1) \cdot t \cdot \tau_0^{1-m} + (\tau_0 - \tau_r)^{1-m}\right)^{\frac{m}{1-m}} \cdot k_1 \cdot (m-1) \cdot \tau_0^{1-m} \qquad (194)$$

the following *limit* is obtained for the slope:

$$\lim_{t\longrightarrow\infty} \frac{d\tau}{dt} = \lim_{t\longrightarrow\infty} \left(\frac{1}{1-m}\left[k_1 \cdot (m-1) \cdot t \cdot \tau_0^{1-m} + (\tau_0 - \tau_r)^{1-m}\right]^{\frac{m}{1-m}} \cdot k_1 \cdot (m-1) \cdot \tau_0^{1-m}\right) = 0 \qquad (195)$$

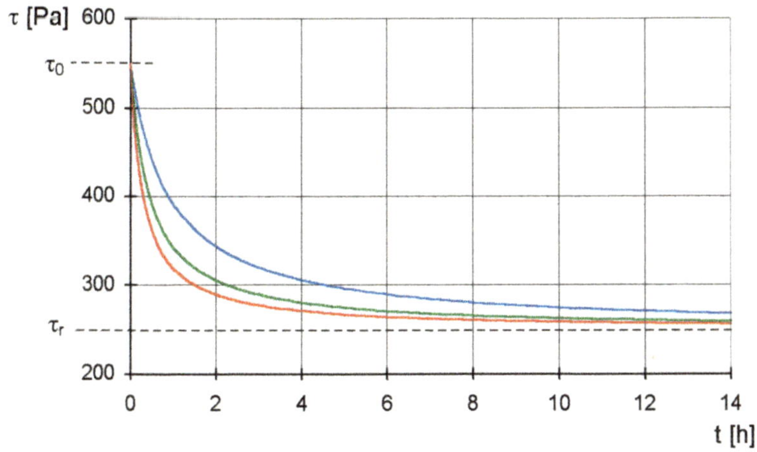

Figure 61: Behavior of the function (192) for different values of k_1 (from bottom to top $6h^{-1}$, $4h^{-1}$, $2h^{-1}$) . Assumed input parameters $\tau_0 = 550Pa$, $\tau_r = 250Pa$, $m = 2$,

Spiegel et al. Equation

The model developed by SPIEGEL ET AL. [159] additionally considers an applied shear rate, with the duration of the stress described by a load cycle number Z. Based on analogical considerations to the WÖHLER-CURVE, [159] assumes the following equation:

$$\tau_f = \tau_{f,\infty} + (\tau_{f,0} - \tau_{f,\infty}) \cdot e^{-\frac{Z}{Z_0}} \tag{196}$$

For the general case of time-varying shear rate $\dot{\gamma}$, it follows

$$Z = \frac{1}{\pi} \cdot \int_0^t \dot{\gamma}(t)dt \tag{197}$$

Where,

t = Duration of stress [s], t_0 = Initial time of stress [s], Z = Load cycle number, $\dot{\gamma}$ = Shear rate $[s^{-1}]$

By substituting Equation (196) into the CASSON-equation (Eq.120), the following relationship is obtained:

$$\left(\frac{\tau}{\tau_{f,\infty}}\right)^n = \left(1 + \left[\frac{\tau_{f,0}}{\tau_{f,\infty}} - 1\right] \cdot e^{-\frac{Z}{Z_0}}\right)^n + \left[\frac{\eta_s}{\tau_{f,\infty}} \cdot \dot{\gamma}\right]^n \tag{198}$$

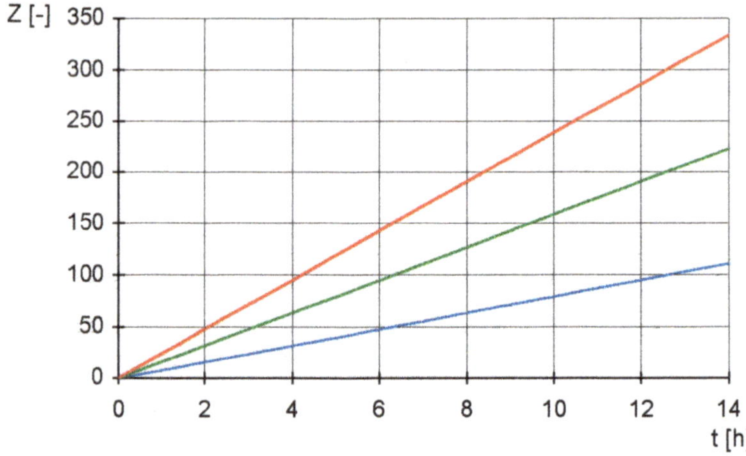

Figure 62: Increase in the load cycle number Z (Eq.197) for different $\dot{\gamma}$ ($25s^{-1}$, $50s^{-1}$, $75s^{-1}$); assuming $t = 0$

for $n < 1$. Where

$\tau(t)$= Shear stress [Pa]; $\tau_{f,\infty}$= Yield stress for $Z \rightarrow Z_0$[Pa]; $\tau_{f,\infty}$= Yield stress for $Z \rightarrow \infty$ [Pa]; Z_0= Reference load cycle number;
n= dimensionless exponent; η_s= viscosity of the grease modified due to thickener (relative to the base oil) [Pas]

To obtain values for the parameters Z_0, $\tau_{f,0}$, $\tau_{f,\infty}$, and n, Equation (198) is adjusted to experimentally determined $\tau - t$ curves. In [159], Spiegel et al. suggest determining the parameter η_s from the slope of the tangent to the flow curve for $\dot{\gamma} \rightarrow \infty$.

Similar to the preceding presentations, Equation (198) is examined for the following conditions:

- Constant shear rate $\dot{\gamma}$ for the entire duration of stress
- Shear rate $\dot{\gamma}$ increasing or decreasing proportionally with time

Constant Shear Rate

The load cycle number then results from the integration of Equation (197) with

$$Z = \frac{\dot{\gamma}}{\pi} \cdot (t - t_0) \tag{199}$$

By substituting into Equation (198) and subsequent rearrangement, the shear stress τ follows as

$$\tau(t) = \left(\left[\tau_{f,\infty} + (\tau_{f,0} - \tau_{f,\infty}) \cdot e^{-\frac{\dot{\gamma}\cdot(t-t_0)}{\pi\cdot Z_0}}\right]^n + (\eta_s \cdot \dot{\gamma})^n\right)^{\frac{1}{n}} \tag{200}$$

Limit

For $t \to \infty$, we have

$$\lim_{t\to\infty} \tau(t) = \lim_{t\to\infty} \left(\left[\tau_{f,\infty} + (\tau_{f,0} - \tau_{f,\infty}) \cdot e^{-\frac{\dot{\gamma}\cdot(t-t_0)}{\pi\cdot Z_0}}\right]^n + (\eta_s \cdot \dot{\gamma})^n\right)^{\frac{1}{n}} \tag{201}$$

thus

$$\lim_{t\to\infty} \tau(t) = \left(\tau_{f,\infty}^n + [\eta_s \cdot \dot{\gamma}]^n\right)^{\frac{1}{n}} \tag{202}$$

Upon reaching this limit, the maximum structural breakdown is also achieved. The described asymptotic stress value corresponds to the previously mentioned equilibrium or residual shear stress. It is evident that the calculated stress is greater the smaller the exponent n and the larger the shear rate $\dot{\gamma}$.

Equation (200) is defined for $t = 0$ and yields

$$\tau(t=0) = \left(\left[\tau_{f,\infty} + (\tau_{f,0} - \tau_{f,\infty}) \cdot e^{\frac{\dot{\gamma}\cdot t_0}{\pi\cdot Z_0}}\right)^n + (\eta_s \cdot \dot{\gamma})^n\right]^{\frac{1}{n}} \tag{203}$$

and for t_0=0

$$\tau(t=0) = \left(\tau_{f,0}^n + (\eta_s \cdot \dot{\gamma})^n\right)^{\frac{1}{n}} \tag{204}$$

Proportional to time-increasing or decreasing shear rate

A time-varying shear rate can be described by the following representation:

$$d\dot{\gamma}(t) = \ddot{\gamma} \cdot dt \tag{205}$$

and upon integration, we obtain

$$\dot{\gamma}(t) = \ddot{\gamma} \cdot (t - t_0) + \dot{\gamma}_0 \tag{206}$$

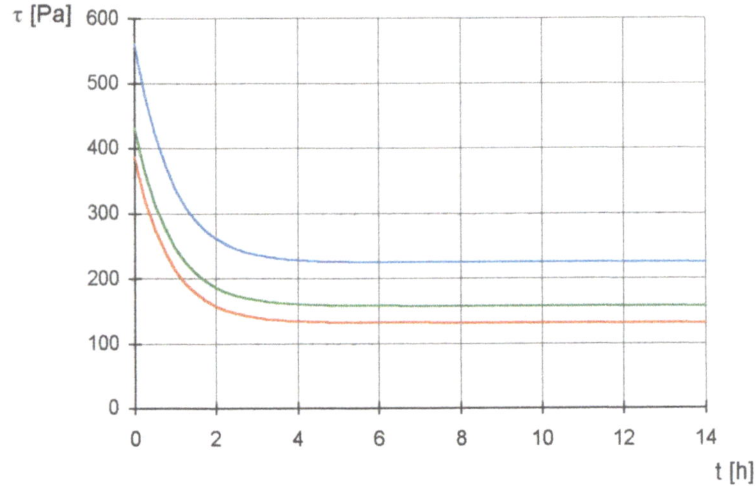

Figure 63: Behavior of Eq. (200) with varying exponent n (0.9; 0.7; 0.5); assuming $\tau_{f,0} = 350\,\text{Pa}$, $\tau_{f,\infty} = 100\,\text{Pa}$, $\eta_S = 0.5\,\text{Pa}\cdot s$, $Z_0 = 50000$, $\dot{\gamma} = 50\,s^{-1}$, $t = 0$,

where

$\dot{\gamma}(t)$ = Shear rate at time t $[s^{-1}]$; $\ddot{\gamma}$ = constant rate of change of velocity $[s^{-2}]$; t = time [s]; t_0 = reference time [s]; $\dot{\gamma}_0$ = reference shear rate $[s^{-1}]$

The relation (206) is then substituted into equation (197), yielding

$$Z = \frac{1}{\pi} \cdot \int_{t_0}^{t} [\ddot{\gamma} \cdot (t - t_0) + \dot{\gamma}_0]\, dt \tag{207}$$

Integration results in

$$Z = \frac{1}{\pi} \cdot \left(\frac{\ddot{\gamma}}{2} \cdot (t^2 - t_0^2) - \ddot{\gamma} \cdot t_0 \cdot (t - t_0) + \dot{\gamma}_0 \cdot (t - t_0) \right) \tag{208}$$

Now, Equation (206) is inserted into Equation (198) and subsequently rearranged, yielding the shear stress:

$$\tau(t) = \left(\left[\tau_{f,\infty} + (\tau_{f,0} - \tau_{f,\infty}) \cdot e^{-\frac{Z}{Z_0}} \right]^n + (\eta_s \cdot \ddot{\gamma} \cdot (t - t_0) + \dot{\gamma}_0 \right)^{\frac{1}{n}} \tag{209}$$

The calculation of the number of load cycles Z in Equation (209) is performed using Equation (207). Equation (209) now provides the opportunity to depict the time-dependent shear stress profile for increasing and decreasing shear rates. Here, the parameters t_0 and $\dot{\gamma}_0$ describe the onset of the change in shear rate.

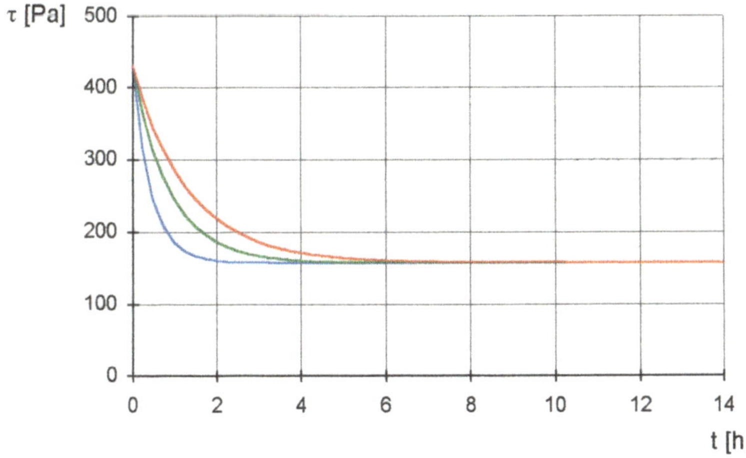

Figure 64: Examination of Eq. (200) for different values of Z_0 (25000, 50000, 75000); assuming $\tau_{f,0} = 350\,\text{Pa}$, $\tau_{f,\infty} = 100\,\text{Pa}$, $\eta_S = 0.5\,\text{Pa} \cdot s$, $n = 0.7$, $\dot{\gamma} = 50\,s^{-1}$, $t = 0$,

Own Empirical Approach

For application in an energetic consideration, an empirical function is developed [160] that reflects the behavior of the lubricating grease and integrates well into tribological analysis.

$$\tau(t) = \tau_{lim} \cdot \left(\frac{t}{t_{lim}}\right)^{-n} \tag{210}$$

Here's the explanation of the symbols:

$\tau(t)$= Shear stress [Pa]; τ_{lim} = Residual shear stress [Pa]; t = Time under load [s]; t_{lim} = Time under load until reaching τ_{lim} [s]; n = Dimensionless exponent describing the intensity of structural breakdown [-]

The following explanation is helpful. The parameters τ_{lim} and t_{lim} describe the state of maximum structural breakdown and form a pair of values on the flow curve. That is, the residual shear stress does not represent an asymptotic value towards which the curve tends for $t \to \infty$.

The *limit value analysis* yields

$$\lim_{t \to 0} \tau(t) = \lim_{t \to 0} \tau_{lim} \cdot \left(\frac{t}{t_{lim}}\right)^{-n} = \infty \tag{211}$$

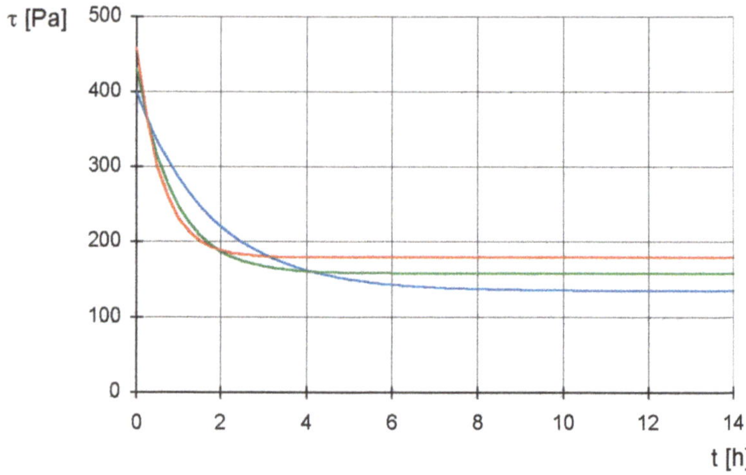

Figure 65: Behavior of Eq. (200) for different $\dot{\gamma}$ values ($25\,s^{-1}$, $50\,s^{-1}$, $75\,s^{-1}$) at $t = 14\,h$; assuming $\tau_{f,0} = 350\,\text{Pa}$, $\tau_{f,\infty} = 100\,\text{Pa}$, $\eta_S = 0.5\,\text{Pa} \cdot s$, $n = 0.7$, $Z_0 = 50000$, $t = 0$

Thus, according to Equation (210), the shear stress tends to infinity as $t \to 0$, which does not seem realistic. This relationship is not defined for $t = 0$.

$$\lim_{t \to \infty} \tau(t) = \lim_{t \to \infty} \tau_{lim} \cdot \left(\frac{t}{t_{lim}}\right)^{-n} = 0 \tag{212}$$

This means that the shear stress decreases below τ_{lim} as $t \to \infty$, which does not correspond to the defined expectations of structural breakdown over time under load.

Equation (210) is defined for $t = t_{lim}$, resulting in

$$\tau\left(t = t_{lim}\right) = \tau_{lim} \tag{213}$$

From these considerations, the argument range of the function (210) is $0 < t < t_{lim}$.

Slope

The slope is given by

$$\frac{d\tau}{dt} = -\frac{\tau_{lim}}{t_{lim}} \cdot n \cdot \left(\frac{t}{t_{lim}}\right)^{-n-1} \tag{214}$$

Considering the point $(t_{lim}; \tau_{lim})$, the equation (210) yields

$$\left(\frac{d\tau}{dt}\right)_{t=t_{lim}} = -\frac{\tau_{lim}}{t_{lim}} \cdot n \tag{215}$$

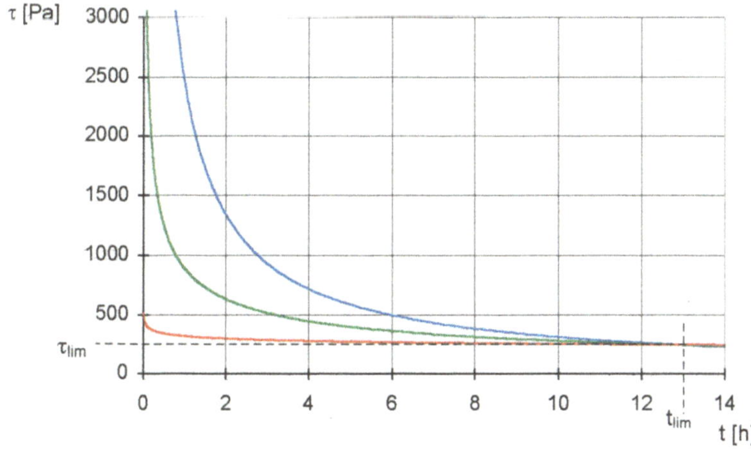

Figure 66: Behavior of the function (210) for different n (0.1, 0.5, 0.9); assuming $\tau_{lim} =$ 250 Pa, $t_{lim} = 13$ h

This means that at the equilibrium state, the slope can take a value $\neq 0$. The slope tends to zero for $t \to \infty$, and we have

$$\lim_{t\to\infty} \frac{d\tau}{dt} = \lim_{t\to\infty} \left[-\frac{\tau_{lim}}{t_{lim}} \cdot n \cdot \left(\frac{t}{t_{lim}}\right)^{-n-1} \right] = 0 \tag{216}$$

Comparison of Time-dependent Models In [146], a numerical regression analysis is conducted to examine the applicability of time-dependent rheological models for describing a long-term experiment. The sum of squared errors S_x is evaluated as a measure of the goodness of fit.

The examination of S_x from the individual analyses yields the order presented in Table 7.1.

7.2.3 Remarks on Rheometry in the Investigation of Lubricating Greases

To describe the flow and deformation behavior of very different substances (*rheology*), experimental work is indispensable. This experimental determination of rheological data (*rheometry*) then provides information to work with developed approaches, narrow down theoretical proposals, or directly compare substances.

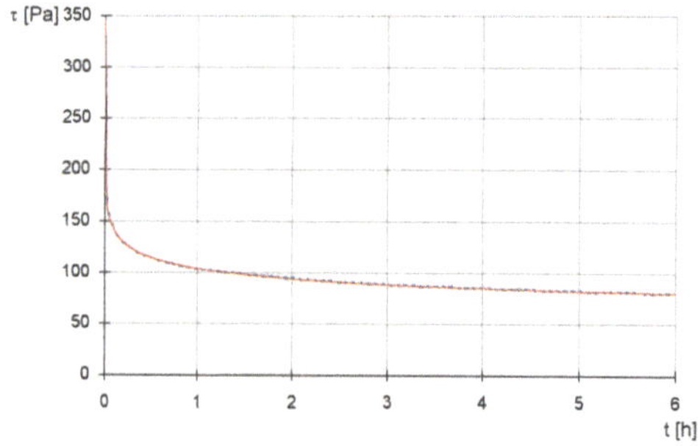

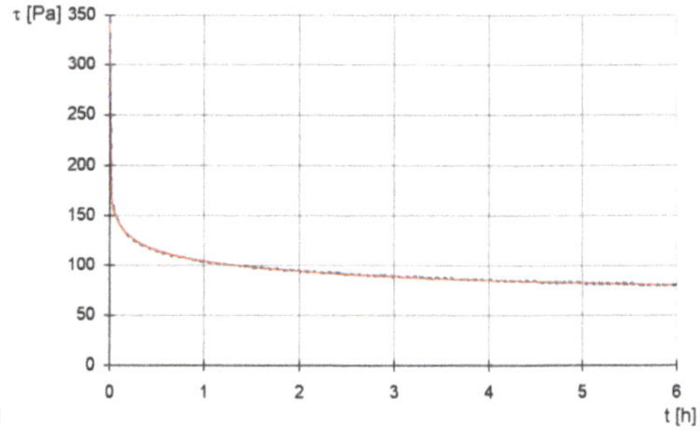

Figure 67: Regression curves for BAUER and CZARNY

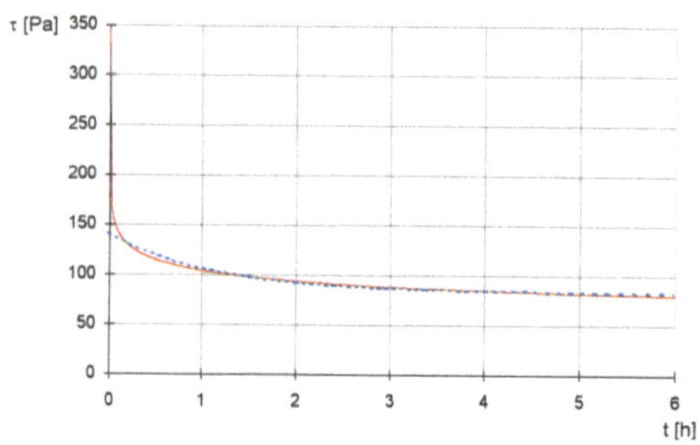

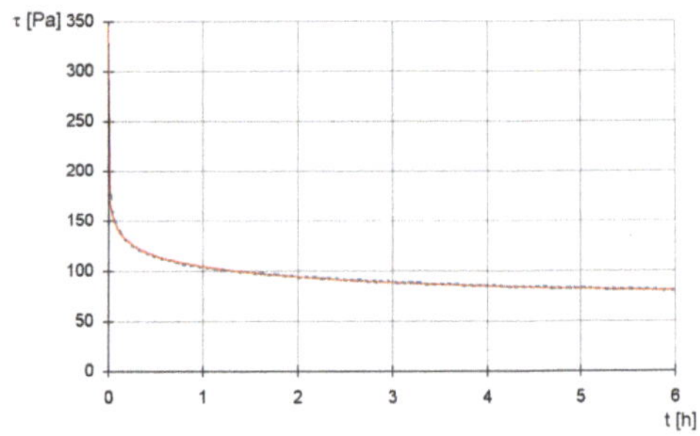

Figure 68: Regression curves for SPIEGEL ET AL. and KUHN

Equation by	Sum of Squared Errors
CZARNY Equation	$S_x = 3.553 \times 10^{-3}$
BAUER Equation	$S_x = 3.556 \times 10^{-3}$
SPIEGEL ET AL. Equation	$S_x = 7.255 \times 10^{-1}$
KUHN Equation	$S_x = 4.739 \times 10^{-3}$

Table 7.1: Comparison of the selected models (dashed line in Figures 67 and 68) regarding the fit to measurement curves

7 Rheological behavior of Lubricating greases

Knowledge of the rheological properties of lubricating greases and their changes due to frictional stress is a prerequisite for treating frictional energy and wear in the grease film. This is a point to which I have already drawn attention at an earlier stage [161], [162], [163], [164].

Lubricating greases are non-Newtonian fluids, as previously explained. The term "fluid"is adopted here due to its traditional use. The peculiarity compared to lubricating oils is a pronounced *viscoelastic behavior*. This means that the behavior of the grease lies between that of an ideal liquid (viscous) and an ideal solid (elastic).

Rheometers are used to determine the rheological properties of lubricating grease samples, which generally work with a plate-plate or cone-plate system when examining viscoelastic substances. In principle, these measuring systems can operate in rotational or oscillatory mode.

Rotation Measurements When observing the viscosity or shear stress behavior over time of lubricating grease samples, rotation experiments with constant shear rate ($\dot{\gamma} =$ const.) are conducted. The modeling and interpretation of these results, also from the perspective of tribological tasks, provide interesting insights for a deeper understanding of system response. In these rheological experiments , the following problems influencing the measurement result should be noted: [165], [166], [167]

- Occurrence of wall slip effects at the solid surface interfaces
- Material extrusion from the measurement gap during the experiment
- Occurrence of negative normal forces at the beginning of shear stress due to deformation of the sample in the gap before the start of the experiment

A typical shear stress - time curve for a lubricating grease (at small $\dot{\gamma}$ values) can be seen in Figure 69.

The examination of the shear stress maximum occurring during the rotation experiment (stress overshoot [165]) for selected model greases is investigated by M. Delgado [168]. There, $\tau_{\max}$ is considered as the shear stress at the transition from (predominantly) elastic deformation to structural change (lubricating grease wear) (see Figure 70).

This study was conducted for a series of selected model greases. For all these model substances, $\tau_{\max}$ approaches an asymptotic value when considering the shear rate gradient $\dot{\gamma}$ (see Figure 71). Detailed information about the model substances and the varied manufacturing parameters are given in [168].

This study was conducted for a series of selected grease samples. For all these model substances, $\tau_{\max}$ tends toward an asymptotic value when considering the shear rate gradient $\dot{\gamma}$ (Figure 71). Detailed information about the model substances and the varied manufacturing parameters are provided in [168].

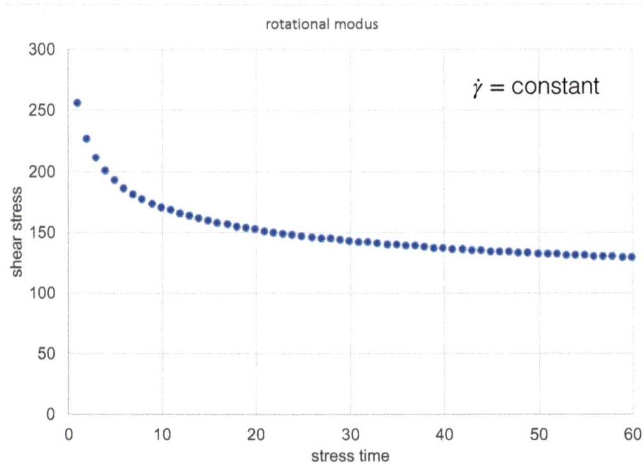

Figure 69: Representation of the typical behavior of shear stress over time for a constant shear rate in the investigation of model greases (representation starts with $\tau_{\max}$)

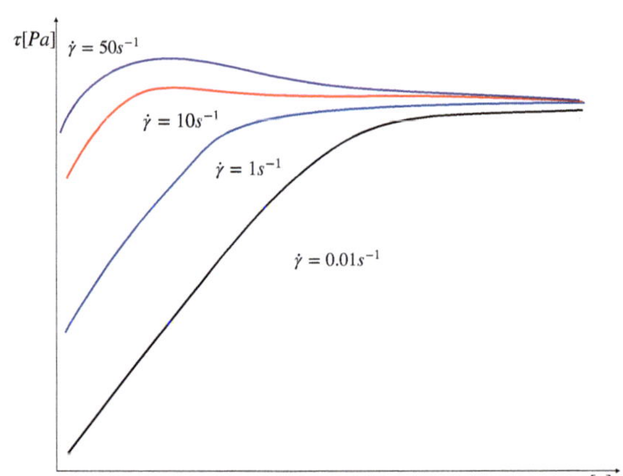

Figure 70: Investigation of shear stress maxima (τ over time) for constant shear rate for a model grease [168] modeled after M. Delgado

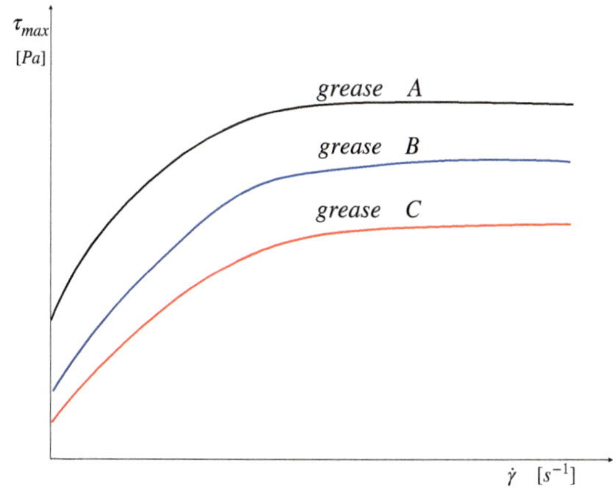

Figure 71: Behavior of $\tau_{\max}$ with respect to shear rate for investigated model greases [168].

Oscillation Measurements To investigate the viscoelastic properties of lubricating greases, oscillation measurements are necessary. A plate-plate or plate-cone measurement system is also used in this case. Other measurements may include creep tests or relaxation tests. Experimental work for determining viscoelastic parameters (oscillation test) is performed in the so-called "Linear ViscoElastic (LVE)"region. This means that the oscillating stress applied does not change the structure under investigation, and, for example, the functions $G'(\gamma)$ and $G''(\gamma)$ remain constant. The deflection of the sample in the measurement gap is illustrated by Figure 72. The vertical dashed line corresponds to the deflection angles of 0° and 180° of an applied oscillating stress ($\pm F$). In the sketch, h represents the measurement gap height, φ is the deflection angle, and s is the deflection.

Performing oscillation with an *ideally elastic* body, the $\tau(t)$ and $\gamma(t)$ curves are always in phase within the LVE region. The temporal change in deformation $\dot{\gamma}(t)$ is shifted by 90°. Considering an *ideally viscous* body, within the LVE region, the $\tau(t)$ and $\dot{\gamma}(t)$ curves are in phase, and $\gamma(t)$ is shifted by 90°.

The behavior of the investigated model lubricating grease will lie "between"the ideal elastic and ideal viscous properties. For the resulting phase shift angle δ, one obtains $0° \leq \delta \leq 90°$. From a tribological perspective, the storage modulus (G'), the loss modulus (G''), and the loss factor ($\tan\delta$ with $\tan\delta = \frac{G''}{G'}$) can be of particular interest. With

$$G' = \left(\frac{\tau_A}{\gamma_A}\right) \cdot \cos\delta \tag{217}$$

and

$$G'' = \left(\frac{\tau_A}{\gamma_A}\right) \cdot \sin\delta \tag{218}$$

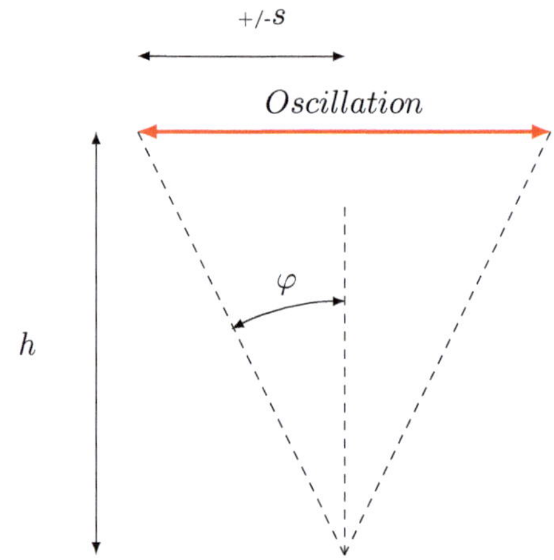

Figure 72: Model representation of oscillation test

The storage modulus $G^{'}$ indicates the ability to store deformation energy, thus characterizing the elastic behavior of the substance under investigation. In contrast, the loss modulus $G^{''}$ represents the dissipated applied energy and therefore describes the viscous behavior. The representation of the complex shear modulus G^{*} using the storage modulus and the loss modulus is shown in Figure 73.

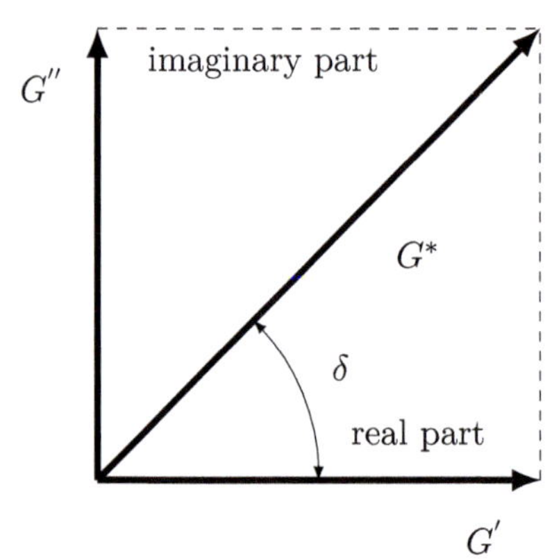

Figure 73: Complex shear modulus with imaginary and real parts

Use of Amplitude Sweep for Tribological Interpretation The so-called *amplitude sweep* is a standard measurement protocol for identifying the linear viscoelastic region (LVE region). For example, a linear ramp of deformation γ is applied at a constant frequency (e.g., 1Hz), while recording the storage and loss moduli. The range where the rheometer records constant moduli (G' and G'' = const.) is designated as the linear viscoelastic region, and dominant elastic deformation is assumed. In the subsequent deformation range, the moduli change, indicating the onset of increasingly plastic deformation. The test continues until the crossover point of the two moduli, where elastic and viscous behaviors are balanced.

With the beginning of the declining storage modulus, a partial flow process begins. As noted in the explanations related to Figure (74), the yield stress $\tau_{Y(osc)}$ indicated by the rheometer is located at this point. The well-known interpretation of the crossover point suggests flow of the entire sample volume. This interpretation appears to be a general convention.

From the perspective considered in this book, the LVE region can be associated with a structure (primary structure) and the subsequent region with a changing structure (formation of a secondary structure) as shown in Figure (74). The process of lubricating grease wear has begun. The transition deformation γ_{krit} can only be investigated and assigned comparatively using the same investigation equipment. The position of this critical deformation will move towards the ordinate with further technical device development, and it is conceivable that a plastic deformation is triggered partially within the lubricating grease volume immediately after initiating a friction process.

Considering the course of both moduli over the (γ) ramp, our investigations show that the graph of (G', following a logistic function with ($\gamma > \gamma_{LVE}$), has an inflection point with characteristic properties for rate and acceleration. This inflection point directly corresponds to the maximum of the (G'') curve (Figure 74). For the investigated model greases, the following applies:

The inflection point of the (G') function corresponds to the maximum of the (G'') curve, thus ($\gamma_{inflection-point} = \gamma_{maximum}$).

A clear physical interpretation cannot be given. Perhaps the peculiarities of the imposed oscillation and the response oscillation of the agglomerates play an important role here.

For energy expenditures considering the structure, the following can be used:

$$e_{rheo_{el}} = \frac{G' \cdot \gamma^2}{\cos\delta} \tag{219}$$

where $\tan\delta$ is the loss factor with $\tan\delta = \frac{G''}{G'}$.

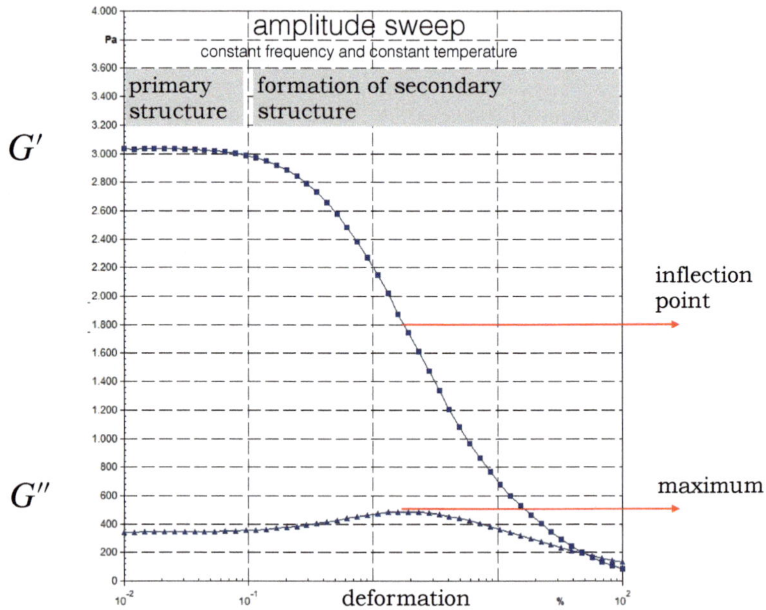

Figure 74: Example of an amplitude sweep for a model grease

In a study by [169] on model greases with different base oils and three solid groups—Li-12-hydroxystearate, diurea, and highly dispersed silica—LITTERS and KOCH demonstrate, among other things, the possibility of correlating the storage modulus G' and the yield stress $\tau_{Y(osc)}$.

They write in their paper: "This demonstrates that elasticity and yield stress in lubricating greases do not necessarily have to be directly related. Although a high polarity in thickener and base oil creates a comparatively elastic superstructure in the short-range order, this quickly collapses under small deformations or spatial displacements of the structural elements. Thus, it becomes less flexible or more rigid."(Figures 75 and 76)

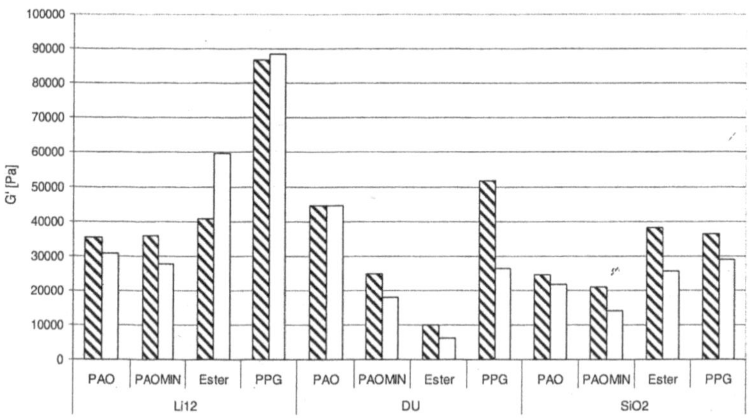

Figure 75: Storage moduli of the grease samples investigated in [169]

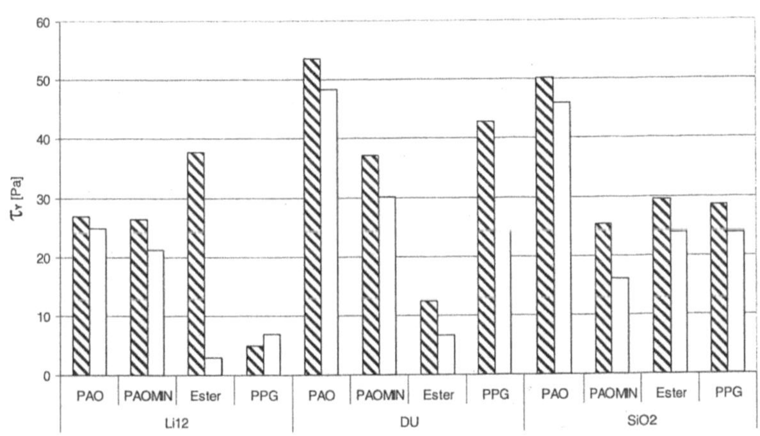

Figure 76: Yield stresses $\tau_{Y(osc)}$ from [169]

8 A Selected Traditional Wear Model

8.1 The Fundamentals According to A. Tross

In the 1960s, ARNOLD TROSS (1893-1969) published a monograph titled *On the Nature and Mechanism of Strength* [170], in which he described deformation and separation processes through the consideration of energy levels, i.e., energy densities.

In this book, he also addresses the friction process and the wear process. For him, (solid-state) wear was associated with the separation of solid-state regions, i.e., reaching the separation energy level. TROSS formulates, for example, flow energy densities, separation energy densities, and fracture energy densities. Besides presenting investigations on an impressive number of material phenomena, he develops the idea that separation processes are triggered when released storage energy excites boundary atoms to the sublimation energy level.

Of particular interest is his *Overlap Theory* , which provides new insights in the description of the states of matter.

Figure 77: Arnold Tross in the 1960s

One finds the following representation by him:

- Materials that are firm but still capable of creep are those whose atoms constantly overlap in all directions.
- Materials that are flowable (slippery) are those that are constantly or periodically freed from their overlaps in individual planes and directions.

- Substances that are liquid are those that periodically free themselves from their overlaps in all directions - whether at individual atoms or molecules or molecular aggregates - and are able to slide past each other in every direction due to their own weight.
- Substances that are gaseous or vaporous are those whose atoms are excited to kinetic energy levels that constantly exceed the binding energy. ([170] p.64)

Furthermore, TROSS writes that in liquids (amorphous state) there is no permanent overlap of the atomic shells. When a liquid is heated, overlaps become smaller or the duration of overlap becomes shorter, and the viscosity decreases.

His concept is illustrated with Figure 78. A more detailed and extensive explanation can be found in the original literature.

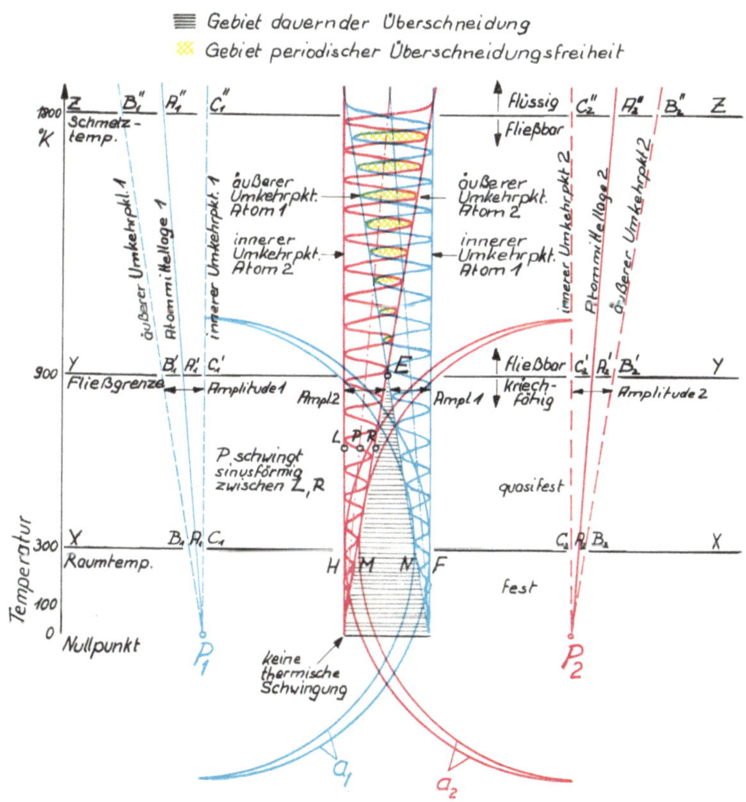

Figure 78: Representation of atomic vibrations about their mean position [170] (Reproduction of the original graph)

In applying the research results of his new strength hypothesis to the friction process and also to the wear process, TROSS provides completely new perspectives in describing the nature of tribological processes.

For instance, he writes about the mechanism of solid-state friction: "he coefficient of friction of solids with cubic or hexagonal normal lattices depends on the force required to perform the work necessary to excite the boundary atoms in the overlap volume to the sliding energy level and thus make the overlapping micro-hills flowable, i.e., plastic deformable, or to shear them by exciting boundary atoms to the separation energy level."Furthermore, he states, "In principle, the laws of plastic deformation apply to friction, and those of separation apply to wear."

To honor his scientific contributions, I have titled the annual Tribology Conference held in Hamburg with his name:

ARNOLD TROSS COLLOQUIUM

In the 1970s and 1980s, G. FLEISCHER, H. GRÖGER, H. THUM and U. WINKELMANN [8] built upon the energetic investigations of TROSS and, with their colleagues, developed a systematic approach for the analytical estimation of solid-state friction and wear.

8.2 Solid-State Friction

TROSS wrote in the 1960s: "According to the plastic deformation hypothesis, the influence of roughness on the coefficient of friction in solid-state friction must depend on the work required to plastic deform the overlap volume."

The systematic works of [8] initially focus on the analysis of friction in contact between rough surfaces. They work with random variables and a deterministic geometry of the microasperities. For a rough-rough contact, the random variables include:

- the relative heights on both contacting bodies ξ_1, ξ_2
- the relative radii ρ_1, ρ_2

The radii belong to the microasperities modeled as spherical segments [171],[101]. The random variables are normally distributed. If both contacting bodies have similar deformation properties, both surface profiles must be considered, and this contact case is referred to as rough-rough. For the case of primary plastic deformation, the real contact area can be written as:

$$A_r = 3.6\frac{r_1 \cdot r_2}{r_1 + r_2}(w_{\text{max}1} + w_{\text{max}2})j_a \int_0^z \int_0^{z-\xi_2} f(\xi_1)f(\xi_2)(z - \xi_1 - \xi_2)d\xi_1 d\xi_2$$
$$\int_{u_k 1}^{u_g 1} f(\rho_1)d\rho_1 \int_{u_k 2}^{u_g 2} f(\rho_2)d\rho_2 \tag{220}$$

Here, the concept in Figure (79) was used. Initially, the formed individual contact area was described with random deformation height, and then multiplied by the contact probability. Here, z is the relative approach with

$$z = \frac{a}{w_{\max}} \tag{221}$$

More detailed information can be found in [172].

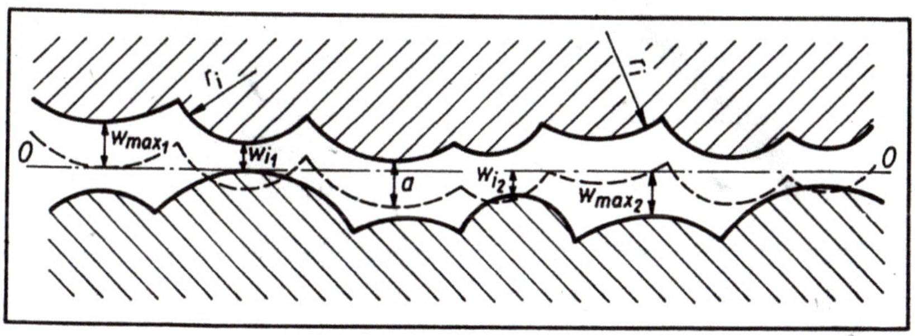

Figure 79: Approximation of two rough surfaces

The energy expenditure occurs at the microcontacts, which, according to the definition formulated here, is identified as friction. One result of the contact geometric investigations is the number of contacts. It is formed from the number of microasperities in a statistically representative section and the contact probability. When determining the energy expenditures, [8] initially consider elastic deformation, plastic deformation, and the separation and deformation of possible adhesive bonds. This is done according to the energy balance of the *dual nature* of friction (Fig. (80)).

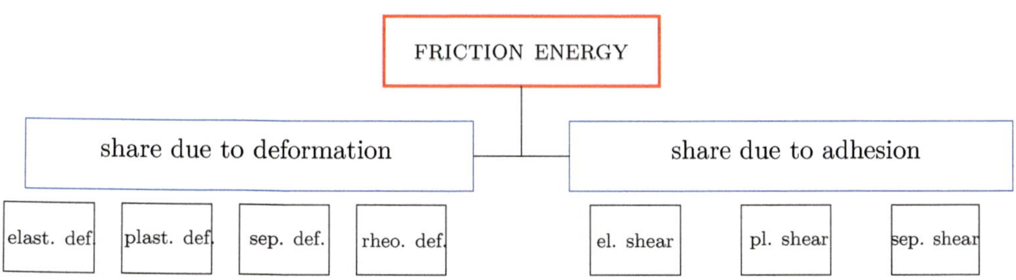

Figure 80: General representation of the dual nature of friction

Based on TROSS, the following mechanisms are considered:

- for elastic deformation
- for plastic deformation
- for the separation of adhesive bonds

and it holds that

$$W_R = \sum_i e_i \cdot n_i \cdot V_i \tag{222}$$

Here, e_i represents the energy density required for each mechanism, n_i is the number of similar energy components in a statistically representative section, and V_i is the volume subjected to friction. Different energy densities have been estimated from macroscopic tests.

- For the energy density required for elastic deformation, the following holds:

$$e_{el} = \frac{\epsilon_V^2 \cdot E}{2} \tag{223}$$

This means estimating the energy expenditure per volume from the stress-strain diagram (uniaxial stress state) [170], [8], [173]. Here, the strain level ϵ_V is defined as

$$0 \leq \epsilon_{Vel} \leq \epsilon_V \tag{224}$$

- For plastic deformation, according to the stress-strain diagram:

$$e_{pl} = \frac{\epsilon_V^2 \cdot E}{2} + \frac{\epsilon_{pl}}{3}(R_e + 2R_m(\psi_B)) \tag{225}$$

and it holds that

$$\epsilon_V \leq \epsilon_{Vpl} \leq \epsilon_B \tag{226}$$

for the range of plastic deformation up to fracture. Here, R_e and $R_m(\psi_B)$ are parameters related to the true stress-strain diagram, considering necking. Studies that define only one random variable in contact modeling [174] describe the strain level as

$$\epsilon_V = \frac{h}{R_Z} \tag{227}$$

where h is the deformation height and R_Z is the mean roughness depth. In modeling with multiple random variables, [172] states

$$\epsilon_{Vij} = \frac{V_1 + V_2}{2 \cdot V_m} \tag{228}$$

The reference measure is a volume of micro-asperities with a mean radius. Every possible contact constellation must be examined.

- When detaching a material particle due to very strong plastic deformation, [175] found

$$e_{sp} = 3.82 \cdot k_{sp} \tag{229}$$

Here, k_{sp} is a parameter from manufacturing technology representing a cutting force.

- To overcome intermolecular bonds, the energy expenditure is given by

$$e_{Tr} = \frac{\tau_S \cdot d_e \cdot \bar{A}_{re}}{V_{Tr}} \tag{230}$$

[8]. From [170] and [174], this mechanism can also be expressed as

$$e_{Tr} = \kappa \cdot c_m \cdot \rho(\vartheta_{subl} - \vartheta_A) \tag{231}$$

Here, τ_S is the shear strength [176], d_e is the mean single contact diameter, $\bar{A}_{re}$ is the mean single contact area, and V_{Tr} is the separation volume. The factor κ (0.3...0.7) accounts for the fact that more than just the boundary atom rows are involved at high plastic deformation.

- For a cutting stress, [177] suggests

$$e_F = \frac{W_R}{V_1} \tag{232}$$

$$V_1 = \frac{V_{Verf}}{\nu_{Verf}} \tag{233}$$

Here, ν_{Verf} is referred to as the consolidation number. Further information can be found in [177]. The study examines the impact penetration of a spherical indenter. The focus is on plastic deformation processes, which are considered as a precursor to abrasive wear. An impactor with a pendulum mechanism was developed for the experimental investigations.

8.3 The Postulate

Even TROSS [170] connects the friction process with wear generation (abrasive wear), and it is likely that the majority of tribologists have observed the close correlation between friction and wear. In the *selected* wear model presented here, the friction process and wear phenomena are linked by introducing a proportionality factor.

Postulate:

$$W_R \sim V_V \tag{234}$$

and further

$$W_R = e_R^* \cdot V_V \tag{235}$$

(in the notation of [8]), where W_R is the friction work and V_V is the worn wear volume.

This formulation applies to the steady state of a wear depth evolution over the friction path. In connection with the wear intensity I_h defined by Kragelski et al. [176], this leads to [63]

$$I_h = \frac{\tau}{e_R^*} \tag{236}$$

This relationship is called the energetic wear basic equation by FLEISCHER [8]. Here, the friction process via the friction shear stress τ is linked to the wear process via the linear wear intensity I_h by means of the proportionality factor e_R^*, the apparent friction energy density .

This system quantity e_R^* is interpreted as the bearable critical energy level of the investigated tribological system. However, in the presentation by [8], it is also bound to other friction states in the solid-body wear (V_V).

The construction of this bearable energy level, i.e., an energy level in the stationary wear phase, is ultimately understood by storing mechanical energy during the repeated contacting of the micro-spots. Afterward, it is shown that

$$e_R^* = \frac{n_k}{\nu_V} \frac{\bar{e}_B}{1 + \zeta_R(n_k - 1)} \tag{237}$$

Central questions were and remain the determination of n_k - the critical number of contacts, and ζ_R - the energy accumulation factor. The critical number of contacts describes the number of contacts until a wear particle is formed, and the energy accumulation factor describes the proportion of energy remaining in the material at each contact. Furthermore, ν_V is a wear rate defined as V_V/V_R, and $\bar{e}_B$ is the mean fracture energy density.

Also helpful in the development of this model were the works of TROSS. He described [170] the mechanism of converting mechanical energy into thermal energy, developed a concept of exciting atoms to higher energy levels, and explained how the stored energy is involved in the separation process.

Investigations into the critical number of contacts and thus into energy accumulation by determining a wear-leading contact proportion and the issue [178]

$$P(W_{rpl} > \bar{W}_B) = 1 - \int_0^{\bar{W}_B} \varphi(W_{rpl}) dW_{rpl} \tag{238}$$

or by applying the cumulative stochastic process in the interpretation as exceeding an energetic limit [179], [180]

$$E(T_b) = \mu_x(1 + H_0(b)) \tag{239}$$

(here, b is an energetic limit and H_0 is the renewal function) provided interesting solutions but were less practical.

The postulate mentioned with Equation (234) as a cause-effect chain is questioned in this book and is reorganized with Figure (9).

9 The Extension of the Wear Concept

9.1 A General Introduction

The definition of wear provided in this work (Definition 2.11) as "the dissipation of frictional energy introduced into a tribological system with simultaneous production of entropy "is in contrast to traditional descriptions of wear, which are limited to the solid body. A system's reaction to frictional stress is attributed to the concept of wear, where material loss is just one possible system response.

Since the formulation of the term Tribology [4] in the mid-1960s, the insights gained in the field of friction and wear research have been analyzed, classified, related, and terminologized in a special way from a systems-theoretical perspective. These efforts undergo constant change with increasing knowledge. This circumstance has led over the years to continuously adapted definitions for many fundamentally observed tribological phenomena. Below, the common concepts of wear are expanded and newly terminologized [13]. In 1960, VOGELPOHL [181], in an effort to complete a given definition for the concept of friction, explicitly excludes the energy expenditure for occurring deformations (in the friction process). He narrows down the wear process to... very slow, gradually progressive changes of surfaces through the abrasion of tiny particles. Also, in a publication from the sixties, TROSS [170] notes that friction is associated with the mechanism of plastic deformation and wear is associated with separation. In the seminal work by KRAGELSKI [176], five types of destructions are formulated from a search. These include elastic displacement, plastic displacement, micro-cutting, changes in adhesion in friction contact, and tearing due to cohesion. Considering the case of a wearless tribological process and a mass change caused by friction on the friction bodies 1 and 2, SADOWSKI [182] discusses in a publication with the description of the *thermodynamic law of wear*. He formulates in [183] "friction is always associated with wear", but also considers the possibility of no wear occurring. From another perspective, FLEISCHER [184] describes material loss or wear as an energetic relief of the tribological system. KOSETZKI [5] formulates stable and unstable regions of the characteristic values of friction and wear for all materials and working media (association of friction and surface destruction). He introduces the ratio of stored energy (G_{Aeff}) to the energy for the formation of secondary structures (G_{SS}) as a criterion for the differentiated states. It is evident that the cited studies elucidate the wear process as a material loss of the friction bodies 1 and/or 2 caused by the effects of the friction process.

The most prevalent general understanding of wear is also expressed in the formulation in the GfT (German Society of Tribology) fact sheet 7 [4]. According to it,"wear is a progressive material loss from the surface of a solid body..."

The effects of friction are now broadened and discussed. KOSJETZKI [5] observes: "As the least disputed result of the entire development of science about friction and wear, the conclusion that external friction is an irreversible thermodynamic process..."

In accordance with PLANCK [185], [62] formulates all natural processes involving friction as irreversible processes, and we recognize precisely in this context the asymmetry of the tribological process. V.WEIZSÄCKER in [186] strengthens this directionality, stating that besides describing the probability that a system not in maximal entropy at time t_1 moves toward a state of higher entropy at time t_2, the application of this probability to the past is excluded. KLAMECKI [34] writes, "The fundamental characteristic of wear is the movement of a body from its relatively well-structured embedding into the environment. The wear process redistributes matter from an initial configuration to a more random one."

It seems reasonable to link these considerations on irreversibility with the investigations into the concept of wear. This entails extending the considerations on the concept of wear to all processes producing this irreversibility.

From these considerations, it is concluded [18]:

Conclusion: *Every application of frictional energy can only be met by the system with irreversible changes*

So, the definition of wear given in this book (Definition (2.11)) describes the general irreversible reactions of the investigated system to frictional stress. This leads to the formulation of an

Impossibility of a wearless tribological process

The subject of investigation is now extended to all system elements of the lubricated friction pair and also to other friction-induced mechanisms, not just material removal.

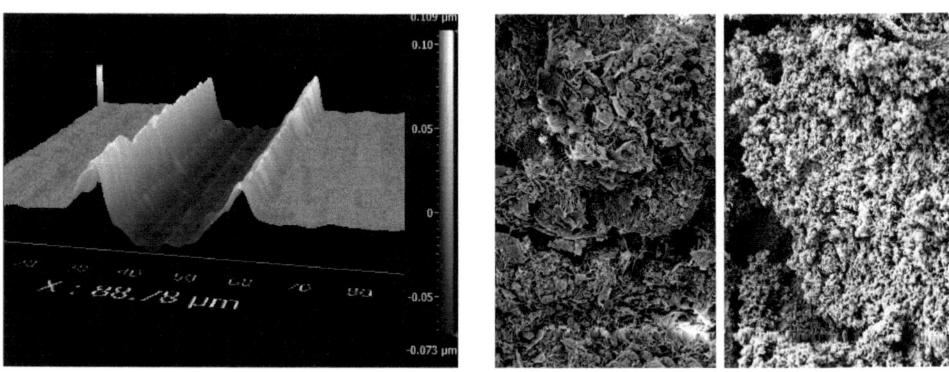

Figure 81: left: Wear track on a steel plate, right: Lubricating grease structure before and after frictional stress

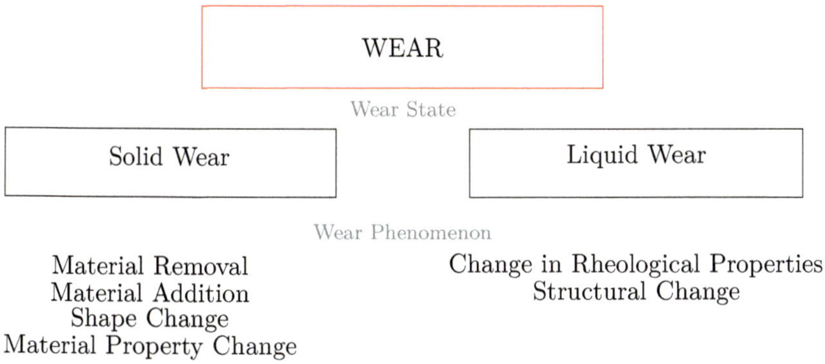

Figure 82: Extended Classification of Wear

For example, the images in Figure (81) are shown.

In a series of experimental studies, the friction effects on model greases were investigated (e.g., [187], [188], [189], [163]). The observable structural changes and property alterations resulting from the introduced friction energy led to the consideration of *lubricant wear* very early on in works such as [20], [19], [190]. Therefore, we can develop the representation shown in Figure (82) within the general discussion. It should be emphasized once again that the logical counterpart to the state of solid wear is the state of liquid wear.

The commonly used terminology such as “removal wear,”“deposit wear,”“shape change wear,”“material property change wear”is contrasted with the term *rheological wear* or *structural wear*. This term describes changes in the rheological properties or the alteration of the original structure as a wear phenomenon.

For the states of wear, the following definitions shall apply:

Definition 9.1 *(Solid Wear) is the dissipation of frictional energy in material regions whose state of matter possesses solid properties.*

and as a terminological compromise:

Definition 9.2 *(Liquid Wear) is the dissipation of frictional energy in material regions whose state of matter possesses liquid properties.*

The starting point for both wear states is an initiated friction process. Although the traditional cause-effect chain is described differently in this book (Chapter (5)), attention is first given to the friction states in the presence of a viscoelastic intermediate substance for this reason.

9.2 Friction in Lubricating Grease

9.2.1 Friction States in the Presence of a Viscoelastic Lubricant

Friction states are classified according to the state of matter of the material regions affected by the friction process [4]. Approaches for the state of solid friction are shown in section (8.2). Now, a viscoelastic lubricant, a lubricating grease, is to be added to the rough microcontact. Subsequently, the state of liquid friction *and* the state of solid friction are possible.

The same terminological objection applies here as with the wear state. Lubricating greases are not liquids. Nevertheless, in tribology, the term *fluid friction* is used for the friction state within the grease film. Even if we consider the base oil of the lubricating grease as a liquid, the term mixed friction (not as a state) and thus the indication of multiple friction states would be correct.

9.2.2 The Grease-Lubricated Contact

For an analytical friction investigation, a contact model should be developed that is based on discrete solid contact (Figure 79) and estimates the energy expenditure at discrete contact points. The contact concept in Figure 79 shows, for example, the deformation of micro-asperities modeled as spherical segments. When a lubricant is added to separate the solid surfaces, the then-imaginary direct contact points become locations of the smallest lubrication gap width. The friction investigations should initially only be conducted at these discrete contact points (Figure 83). If we investigate the microgeometric single

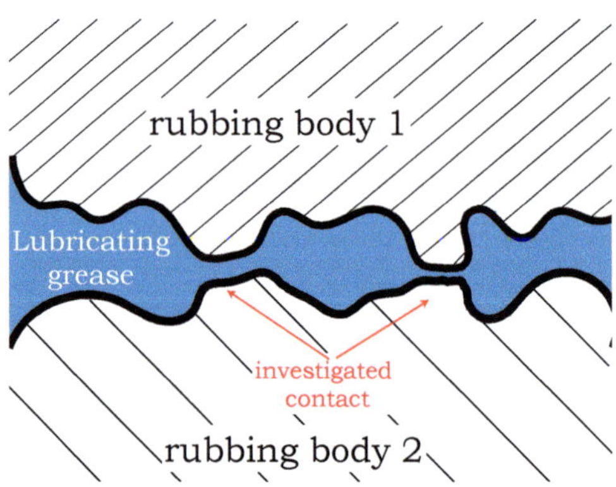

Figure 83: The grease-filled contact with modeled micro-asperities

contact of two micro-asperities modeled as spherical segments, which are separated by the lubricating grease film, the concept shown in Figure (84) should apply. This takes into account the heterogeneity of the lubricating grease through individually differently structured areas (corresponding to the density regions). These different density regions, which also represent different property areas, were statistically investigated for selected model greases. They can be described by the density function of the normal distribution, just like the height of the solid asperities and the radii of the modeled spherical segments. IR-microscopic investigations by W. HOLWEGER [75] were evaluated for this purpose, leading to a topography of the lubricating grease film (Figure (84)). We will now consider

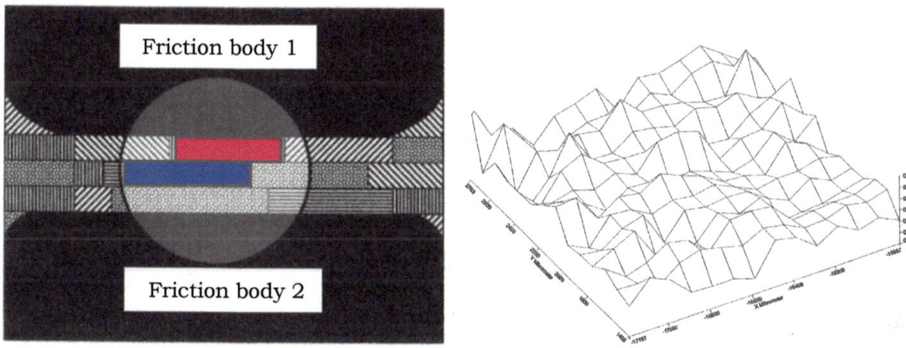

Figure 84: The investigated single contact of the grease-filled lubrication gap and the topography of a lubricating grease

the probability of forming a (described) contact, both between the micro-asperities and between the investigated property regions within the lubricating grease. In a first step, contact formation between two layers in the grease film will be included. This involves working with 6 random variables, as the examined contact consists of two rough surfaces. These random variables are:

1. the heights of the micro-asperities on friction body 1 and friction body 2,
2. the radii of the modeled spherical segments on friction body 1 and friction body 2,
3. the property areas in layer 1 and layer 2.

Typically, relative sizes are used, which include the relative height ξ, relative radius ρ, and relative density δ. For the contact described here, the following applies:

$$F(z,u,\rho) = \int_0^z \int_0^{z-\xi_2} f(\xi_1)f(\xi_2)d\xi_1 d\xi_2 \int_{u_k}^{u_g} f(\rho_1)d\rho_1 \int_{u_k}^{u_g} f(\rho_2)d\rho_2 \int_{\rho_2}^{\rho_1} f(\delta_1)d\delta_1 \int_{\rho_2}^{\rho_1} f(\delta_2)d\delta_2 \tag{240}$$

The integration limits for approaching two rough surfaces were applied according to [191]. They serve as upper and lower bounds for the radii and density regions under consideration.

The analysis assumes independence of heights and radii [171], as well as independence of density regions. The number of contact points (smallest expected gap width and contact of the selected density regions) is then determined by

$$j_r = j_a \cdot \int_0^z \int_0^{z-\xi_2} f(\xi_1) f(\xi_2) d\xi_1 d\xi_2 \int_{u_k}^{u_g} f(\rho_1) d\rho_1 \int_{u_k}^{u_g} f(\rho_2) d\rho_2 \\ \int_{\rho_2}^{\rho_1} f(\delta_1) d\delta_1 \int_{\rho_2}^{\rho_1} f(\delta_2) d\delta_2 \tag{241}$$

Where j_a represents the total number of micro-asperities. When using these equations, calculations are conducted using the standardized normal distribution.

9.2.3 Energy Expenditure in Lubricating Grease

Energy Expenditure in Shearing of Lubricating Grease

As shown in [71], the dissipation function in rheometer tests can be represented as the product of shear stress τ and shear rate (deformation rate) $\dot{\gamma}$. For the energy density expended as an expression of frictional events in the grease film [164], [192], [193], we obtain

$$e_{rheo} = \dot{\gamma}_c \cdot \int_0^{\zeta} \tau(t)\, dt \tag{242}$$

where $\dot{\gamma}_c$ represents the shear rate, ζ denotes the duration of stress, and t is the instantaneous time. For comparative studies, using a rheometer is a suitable method to characterize the friction behavior of different lubricating grease samples (see Figure (85)). In these experiments, the shear rate $\dot{\gamma}$, test temperature ϑ, and ambient pressure p can be varied as needed.

Comparative experiments are conducted under identical test conditions. The shear stress behavior at constant shear rate is depicted in Figure (86).

Here, the area under the τ vs. t curve at constant shear rate represents energy expenditure per volume (Equation (242)). This allows for evaluating the friction behavior in the grease film of different samples under identical stress durations and shear rates.

An empirical suggestion for describing the shear stress variation [160] is

$$\tau(t) = \tau_{lim} \cdot \left(\frac{t}{t_{lim}}\right)^{-n} \tag{243}$$

An example of determined energy densities at different temperatures for shearing a model grease is shown in Figure (87).

Interpreting the curve enables further considerations that will be discussed later in this book.

With

$$e_1 = \dot{\gamma} \cdot \int_0^{t_{max}} \tau(t)\, dt \tag{244}$$

Figure 85: Lubricating grease examination on a rheometer using a plate-plate measurement system

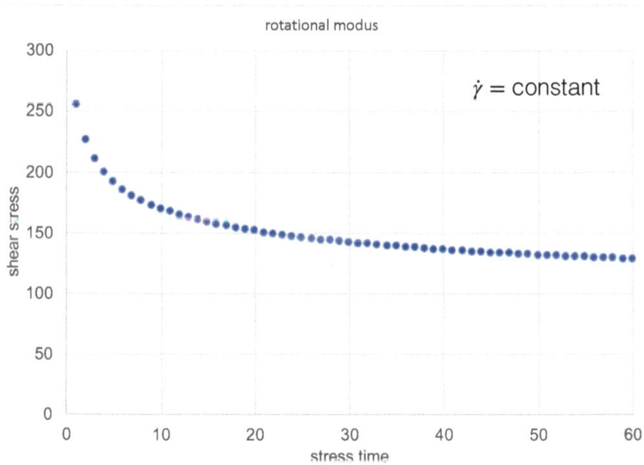

Figure 86: Example of a flow curve of a model grease at constant shear rate

DELGADO et al. [188], [194] describe a stored energy density until reaching τ_{max} for the initial region not shown in Figure (86). Another experimental approach involves oscillation measurements. Comparative friction studies, for example, can be conducted in the region before macroscopic yielding begins. During an amplitude sweep, typical trends of the storage and loss modulus can be observed in lubricating grease investigations (Figure (89)).

For a selected deformation, such as the transition from the *linear viscoelastic region*

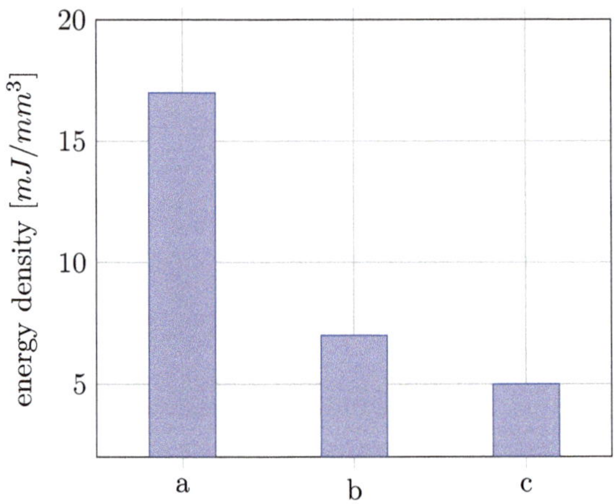

Figure 87: Energy densities in rotational tests with varying test temperatures, with a=25^0C, b=40^0C und c=80^0C

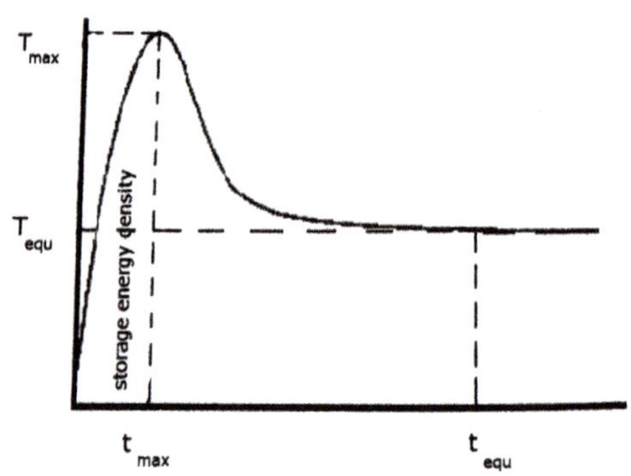

Figure 88: Investigation by DELGADO of the elastic region [188]

to the plastic region (yielding), the influence of the storage modulus, which represents structural effects, on the friction process can be comparatively examined.

It is proposed that

$$e_{Rrheo} = \frac{G' \cdot \gamma^2}{\cos \delta} \quad (245)$$

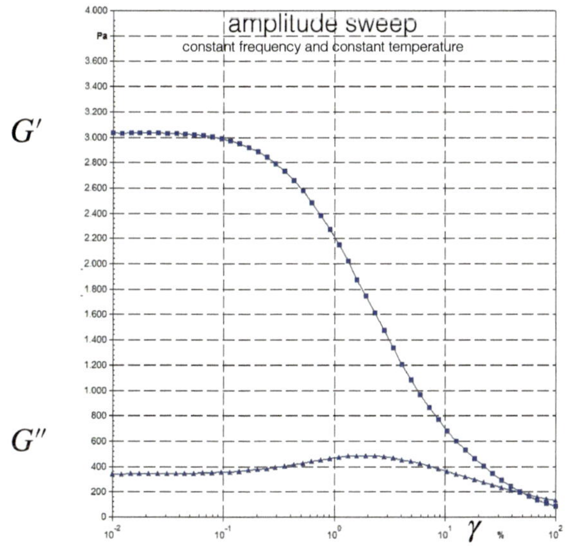

Figure 89: Typical behavior of G' and G'' during an amplitude sweep with a model grease.

Energy Expenditure under Application of Normal Force

An experiment that appears unconventional also provides interesting insights. A compressive force is applied to the sample under investigation. Ultimately, a flow process is triggered, causing the sample to exit the gap. The variable force applied can be observed over the deformation path, meaning the initial gap decreases with increasing compressive force (Figure 90). The energy expenditure during the deformation of a model substance is evaluated.

Figure 90: Rheometer gap before and after the deformation experiment

The typical behavior under application of a force ramp exhibits differentiated stages

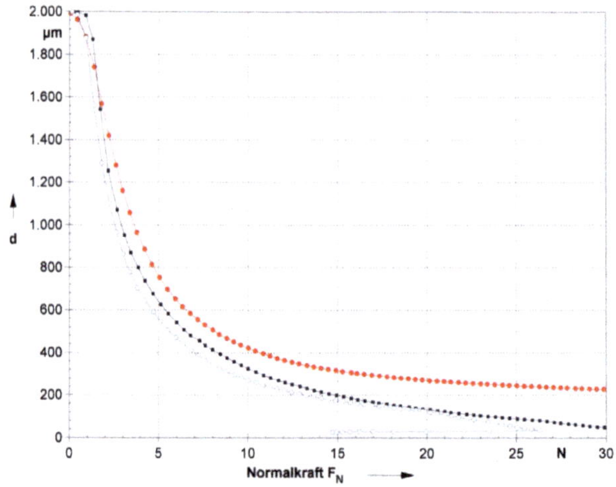

Figure 91: Change in gap width under application of a compressive force ramp 0...30 N for samples with different solid content (10%, 22%)

over the course of the experiment. The following regions can be identified:

- Region I: Initially dominated by elastic deformation.
- Region II: Followed by nearly stationary flow (plastic deformation).
- Region III: A kind of solidification effect that determines a changing ratio of base oil to solid content upon exiting the gap.

An intriguing result emerges when periodically pausing the deformation process to conduct oscillation measurements. Some samples exhibit a significant increase in the storage modulus. This might indicate an uneven expulsion of base oil and solid content.

Referring to Figure (91), we observe the force-displacement curve for samples with different solid contents. The sample with lower solid content requires a higher force from the rheometer for a comparable gap size. Towards the end of the experiment, the storage modulus for this sample was significantly higher compared to the sample with higher solid content. One possible interpretation is that the sample with lower solid content expelled a higher proportion of base oil (indicating a relatively weaker structure), such that by the end of the experiment, it was no longer a 10% solid content sample within the gap (Figure 92). From these deformation experiments, an analogy to solid body behavior allows for the derivation of a stress-strain diagram (compression)(Figure 93).

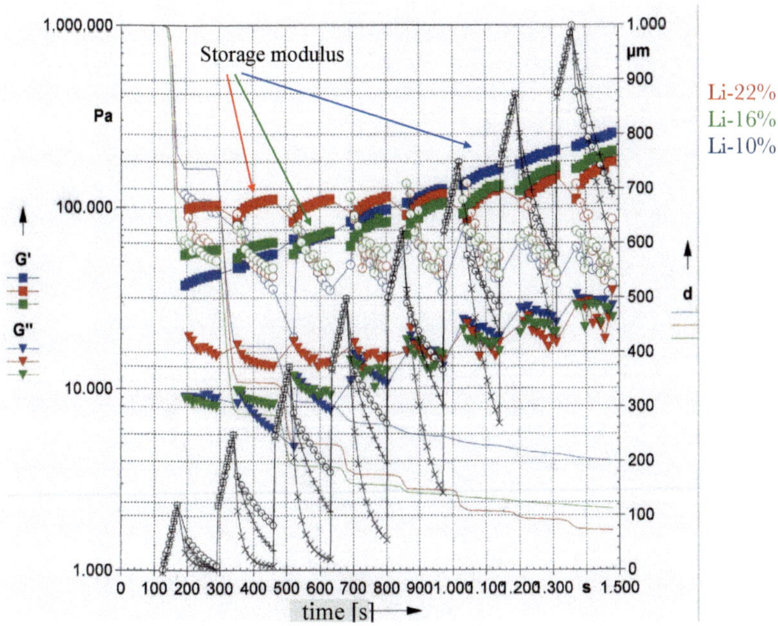

Figure 92: Change in storage modulus during deformation experiment for 3 grease samples

The regions can be described as follows:

Region I (elastic deformation):

$$e_{Nel} = \frac{\sigma_V \cdot \psi_V}{2} \tag{246}$$

Region II (viscous flow):

$$e_{Nvisc} = \sigma_V \cdot (\psi_S - \psi_V) + \frac{\sigma_V \cdot \psi_V}{2} \tag{247}$$

Region III (solidification):

$$e_{Ns} = \frac{A}{B} \cdot (e^{B\psi_0} - e^{B\psi_S}) \tag{248}$$

σ_V and ψ_V denote the beginning of the viscous section. In contrast, σ_S and ψ_S represent the onset of the solidification phase. ψ_0 describes the end of the third section. Parameters A and B are utilized for characterizing the solidification period based on the conducted regression.

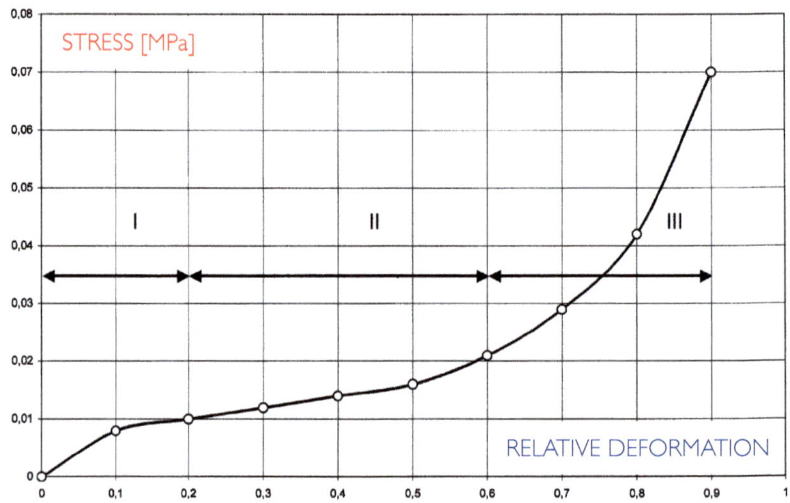

Figure 93: Stress (σ)-strain (ψ) diagram of a grease sample with the aforementioned regions I to III [195]

Possibility for investigating cohesion behavior

The investigation of the cohesion behavior of lubricating greases (as shown some time ago in [195] and [196]) appears to be particularly useful in the context of rolling bearing lubrication and attempts at energy balancing. The rheometer can also be used for these investigations, although with a slightly modified experimental procedure. Essentially, the counterpart to the deformation experiment shown in Figure (90) is performed. In this test, force and displacement are recorded during a tensile test until a lubricating grease thread breaks (Figure 94, Figure 95). Experience shows that cohesion forces are always smaller compared to adhesive effects. An energetic evaluation can also be performed in this experiment [195], [196]. The section up to the inflection point describes the efforts in the elastic region. The inflection point marks the transition into plastic deformation, and a tensile force $F = 0$ indicates that the formed lubricating grease thread has been separated. Investigations have shown that the solid content significantly influences the curve progression. Very similar behavior was observed in experiments with a pendulum tribometer [197]. Recent more comprehensive studies can be found in [198].

Figure 94: Tensile test on the rheometer with a lubricating grease sample

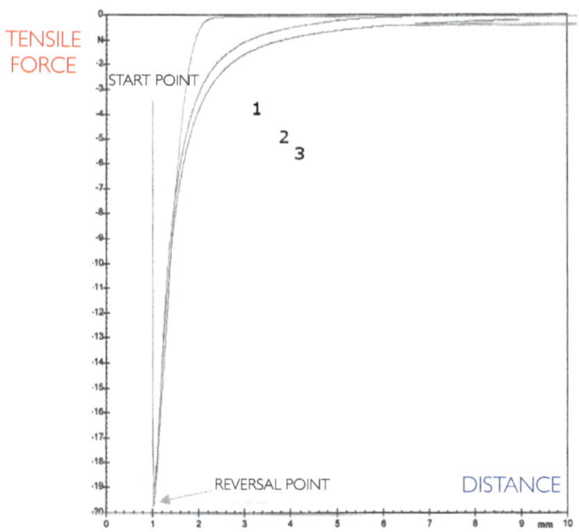

Figure 95: Variation of tensile force over displacement for 3 different samples

Some Remarks on Mixed Friction

There exist various descriptions and definitions for mixed friction. Under the heading *Friction State*, one can read in [4], among other sources: "Mixed friction is called any mixed form of friction states, primarily of solid and liquid friction." On the other hand, [14] defines it as: "Mixed friction is the mixed form of at least two friction states that occur simultaneously and side by side." This definition can also be found categorized under the heading of friction states.

Definition 9.3 *(Mixed Friction) Mixed friction describes the simultaneous and side-by-side occurrence of different friction states. It is not a friction state itself.*

Even though the term is widely used to describe a friction state, this usage is incorrect according to established definitions. A friction state is tied to the phase of a material region. The term *mixed friction* merely indicates that different friction states occur in the considered process section or in the examined contact.

In technical tribological contacts with an intermediary substance, the presence of *mixed friction* is of great importance and is extremely common. In most cases, this refers to the occurrence of both solid friction and liquid friction. Examples include machine elements/assemblies such as rolling bearings, gearboxes, guides, etc. In these cases, both friction states generally occur simultaneously and side by side without direct solid-solid contact. Through the closed lubricant film, surface roughness is deformed, and at the same time, the lubricant, such as oil, is subjected to stress (Figure (96)). The frequently expressed notion of the "squeezing through" of the lubricant film has been refuted by the works of [14], [199], [200], and [201]. The general friction behavior in a lubricated friction

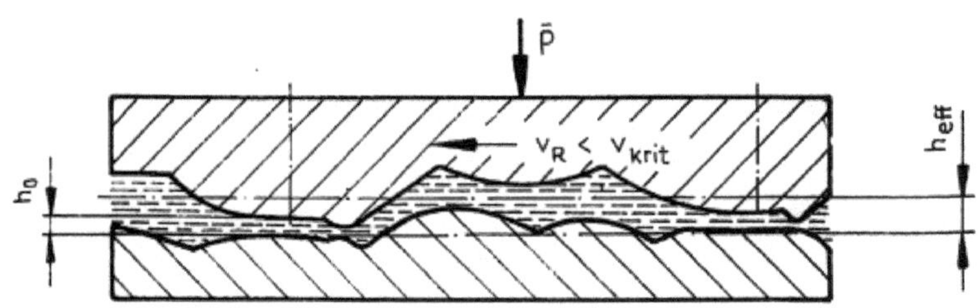

Figure 96: Mixed friction contact—here EHD conditions according to [14]

pair as a function of relative speed and lubricant film thickness is illustrated in Figure (97).

These considerations of mixed friction remain unchanged if the intermediary substance is a lubricating grease, although in highly stressed contacts, the formation of the grease film may be influenced by, for example, soap structure materials [87].

In general, the frictional energy of a mixed friction pair can be expressed as

$$E_{\text{friction}} = E_{\text{solid}} + E_{\text{liquid}} \tag{249}$$

9.3 Lubricating Grease Wear

Lubricating grease is initially treated as an element, a machine element, in terminological terms. Thus, one can speak of lubricating grease wear, similar to gear wear or sliding

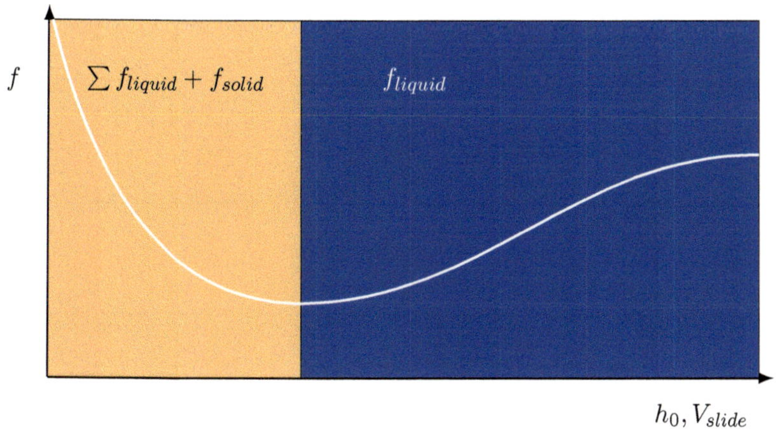

Figure 97: Friction behavior of a lubricated pair (In the first section one often finds the designation mixed friction and for the part before the friction minimum the designation EHL.)

bearing wear, etc. A tribological analysis of the present wear states or types can then only reveal solid wear (fragmentation of agglomerates) and/or liquid wear (frictional effects in the base oil, such as heating).

By analogy with the commonly used, albeit incorrect, term for friction state, mixed friction, one might refer to this as *mixed wear* (see Fig.98).

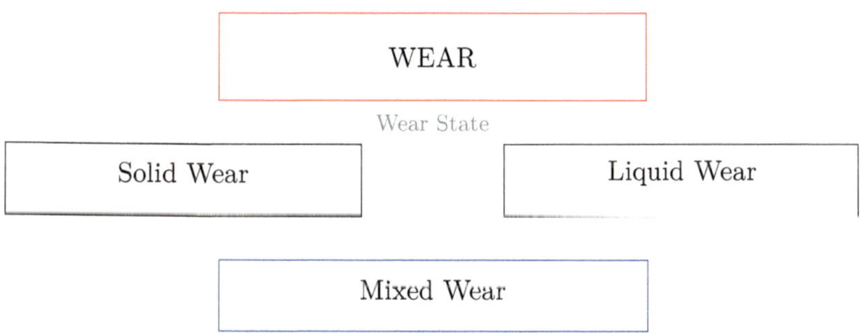

Figure 98: Both wear states can occur simultaneously in a tribological contact. This is described by the term *mixed wear*.

9.3.1 Some Historical Aspects

In a very interesting paper, LIEBL and VAMOS [202] published considerations in 1968 regarding the influence of soap structure on the typical time-dependent flow behavior (η vs. t). In this work, lubricating greases are characterized as phase colloids with a continuous liquid phase and a coarsely dispersed or deformable phase. For the general trend of viscosity η^* over shear stress τ, they provide the following explanation (see Figure 99).

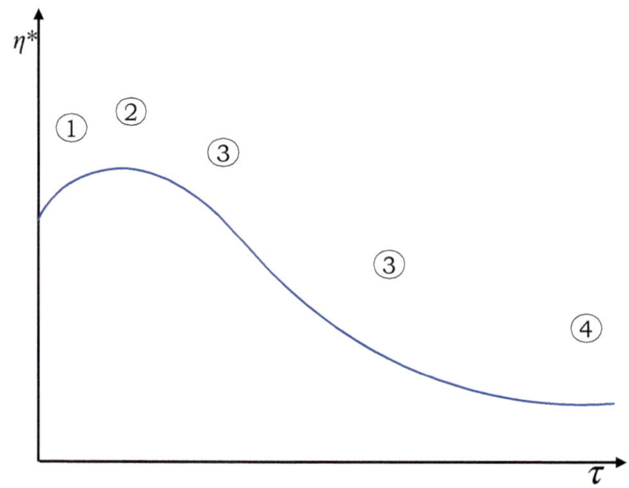

Figure 99: Variation of apparent viscosity with shear stress according to [202]

- 1: Soap fibrils at rest
- 2: Elastic deformation of fibrils
- 3+4: Breakage of fibrils

In other words, the structural changes [202] describe processes such as

- reversible elastic deformation
- reversible orientation
- irreversible dispersion

Against the background of the considerations presented in this book, this phenomenological representation does not address the driving forces of the occurring wear, nor was a thermodynamic analysis the goal. Therefore, the use of the terms reversible and irreversible should be viewed in the context of the observations presented by [202]. CZARNY extensively investigates the time-dependent behavior of lubricating greases. He explains the typical trend of apparent viscosity η^* or shear stress τ over the exposure time t (see also Figure 99) through two opposing mechanisms [189]: "On one hand, the destruction of structural bonds, and on the other hand, the simultaneous (and partial) rebuilding of new bonds." According to CZARNY, a dynamic equilibrium is reached when the rate of decay and the rate of formation are balanced (this state is primarily influenced by $\dot{\gamma}$). Even after prolonged shear periods, certain *limit stresses* [66] are still measured, indicating that the spatial structure is not entirely destroyed.

Work by HOTTEN [203] on changes in grease structure during the friction process in rolling bearings, which he conducted using optical investigation methods, concludes that an equilibrium state is achieved through destruction and rebuilding, although this state is different from the initial one. According to SPIEGEL ET AL. [159] , the decreasing shear stress over time under constant shearing is attributed to the "grinding" of the grease structure. This process tends towards a stable particle size. Additionally, [204] describes that through the rolling of the grease, the thickening structure is gradually broken down.

In earlier editions of this book, such as [205], analogies to solid wear were considered, and a critical energy level e^*_{Rrheo} was described. The concept was that a portion of the applied frictional energy accumulates in the lubricating grease and triggers wear in the volume element after a critical exposure time. The equation was given by

$$e^*_{Rrheo} = \frac{e_z \cdot \vartheta}{\kappa \cdot \nu} \tag{250}$$

where e_z represents a critical energy level of the material that builds up after a critical exposure time t_z:

$$e_z = \dot{e}_{Def} \cdot \kappa \cdot t_z \tag{251}$$

and κ is an energy accumulation factor. Here, ϑ is the ratio of exposure time to critical exposure time, and ν is the ratio of wear volume to friction volume. More detailed information about the critical exposure time using a probabilistic approach can also be found in [205]. Due to the limited practicality of this approach, a more detailed description was omitted. The same applies to the development of a wear contact model when considering level exceedances (see Fig. (100)).

These considerations led to an expected value for the number of level exceedances given by

$$E[N_0(u)] = \frac{1}{2\pi}\sqrt{\frac{m_2}{m_0}} \quad e^{-\frac{u^2}{2m_0}} \tag{252}$$

where N_0 is the number of level exceedances, m represents the spectral moments, and u is the critical energy level. While this model provides further insights, it is similarly impractical.

The extension of the wear concept and its terminological introduction were proposed by me in 1992 with [17] and in 1994 with [20].

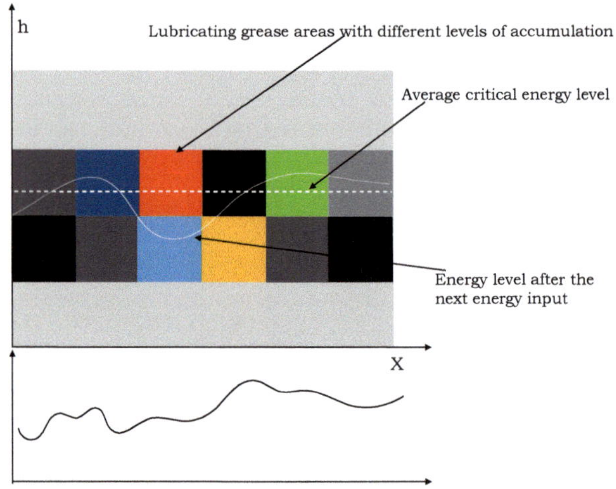

Figure 100: Model presentation for the description of wear formation in the lubricating grease film [206], with h being the energy accumulation level.

9.3.2 Selected Studies on the Friction Effects in Lubricating Grease

Early studies by SPIEGEL ET AL. [159] investigate the behavior of selected lubricating greases regarding structural changes. They conduct loading and unloading ramps and demonstrate the *irreversible* nature of the process. SPIEGEL develops a model of lubricating grease wear, which he refers to as *shear strength*. This involves examining varying shear stresses on lubricating grease particles.

In the work of FRANCO ET AL. [207] (Fig.(101)) , we find quantitative statements obtained from comparing parameters from oscillation tests.

In numerous studies on grease thixotropy, PASZKOWSKI and OSZTYNSKA-JANUS [208], [209] describe the decrease in shear stress over time under constant shear in a rheometer. Although these studies focus on the thixotropic effect, they thoroughly analyze the stress phase. It is shown, among other things, that hydrogen bonds connecting the OH groups are destroyed during shearing.

The investigation using rheometer penetration measurement is undertaken by REZA-SOLTANI [210]. The *net penetration* [*mm*] is measured after different stresses. He shows a linear relationship between stress and net penetration. He also extensively examines the structural breakdown due to chemical processes.

A completely different consideration of grease changes due to frictional stress is shown by KHONSARI ET AL. [211]. They measure, among other things, the contact angle of a water droplet placed on the grease surface. Grease samples from various intensities of stress in a grease worker are used. It is observed that the contact angle tends to decrease with an increasing number of strokes, establishing a correlation with grease wear (Figure

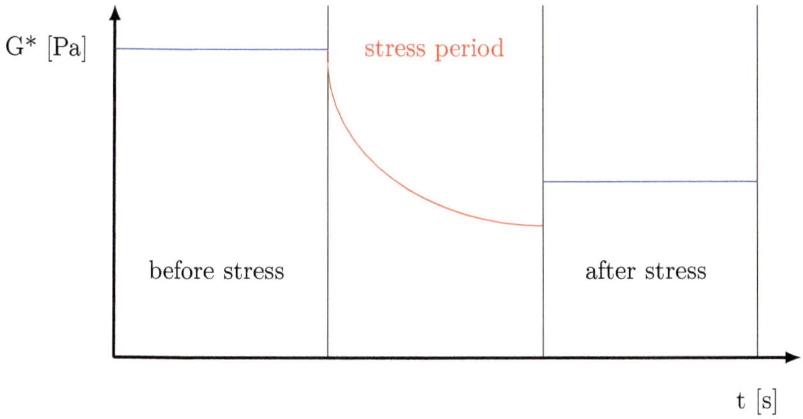

Figure 101: Experimental investigations in the oscillation test through a loading and unloading cycle. After [207]

(102)).

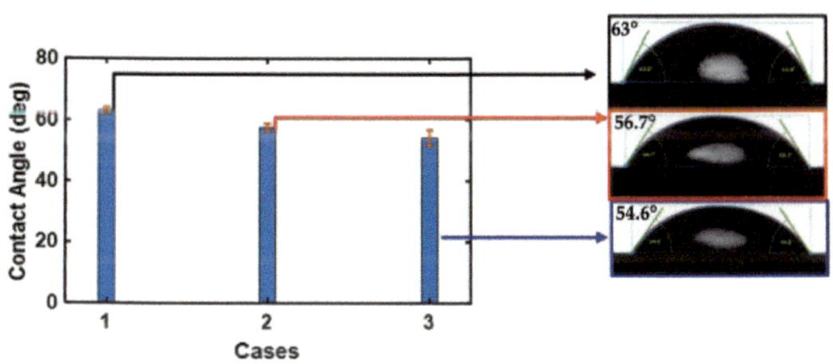

Figure 102: Contact angle as a function of differently intensively worked samples of a PU grease [211]

In a very comprehensive study, Y. Zhou, R. Bosman and P.M. Lugt [97] developed, among other things, a zero-viscosity rate and related it to a produced entropy density. They describe a grease aging equation with

$$Y = \frac{Y_i - Y_\infty}{1 + K \cdot S_g^m} + Y_\infty \tag{253}$$

where Y represents the rheological properties, Y_i the initial values of the rheological parameters for the unstressed grease, Y_∞ the values for long-term stress, S_g the produced

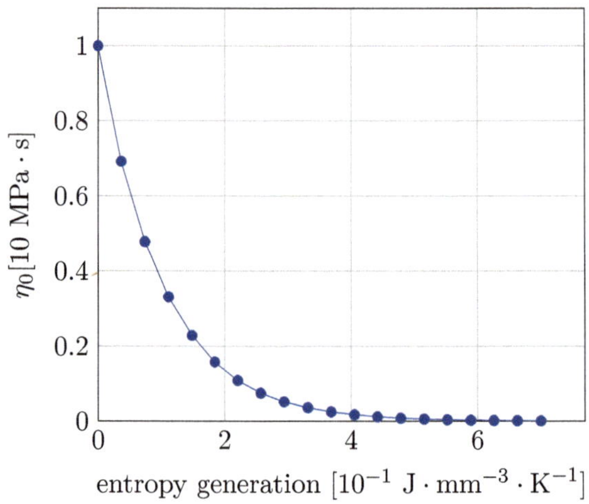

Figure 103: Example representation of the relationship (253) for the zero-viscosity rate

entropy per volume, K the degradation coefficient, and m the degradation exponent. This means that the relationship (253) is applied to different rheological properties, and each sample has its own master curve.

Lijesh et al. [212] describe the *lifetime* of lubricating greases. A newly developed rolling bearing test rig is used to generate grease wear in a real contact. In an adapted rheometer, greases are sheared and then the reduction of the gap is measured with an applied normal force $F_N = 2N$. These experiments provide different values for differently produced grease wear. Figure (104) shows the correlation between squeeze behavior and entropy production. Further information is provided by [213].

Osara and Bryant [214] extensively investigate the behavior of lubricating greases from the perspective of varying energetic states. Their work covers a wide range of traditionally observed phenomena and analyzes the energetics of the processes. Interestingly, they perform temperature measurements with their developed device for simulating grease stress and correlate these with shear stress development, as shown in Figure (105).

9.3.3 Experimental Investigations on Grease Wear

The Rheometer Procedure The defined wear of the grease manifests itself in the change of its solid structure, i.e. the geometry and arrangement or distribution of the agglomerates and fibrils. It represents the system's response to achieve a stable state through an energy-dissipating process. Under specific process conditions, the system may respond to instability with structural formation, thereby adjusting the energy dissipation.

It should be reiterated here that grease wear due to mechanical energy stress is under

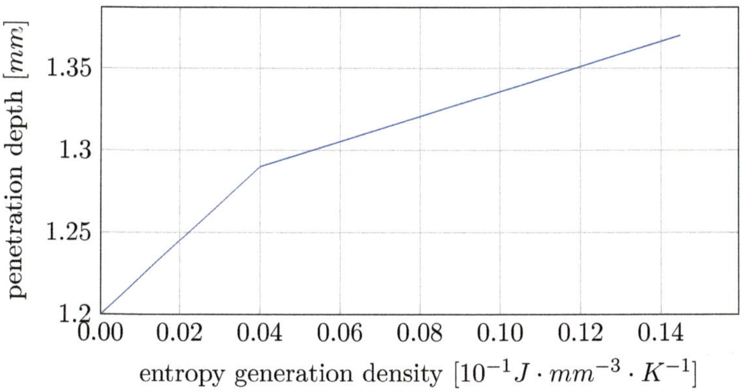

Figure 104: Regression line as an example for investigations according to [212]

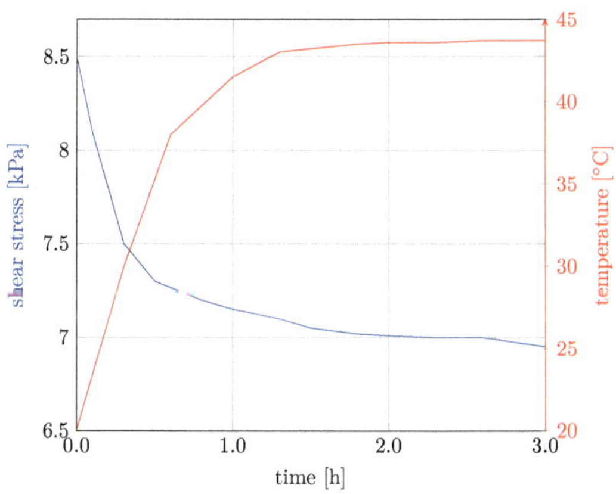

Figure 105: Shear stress and temperature development, exemplarily according to the experiments in [214]

investigation.

Experimentally, the following possibilities exist:

- Structural investigation through before-and-after examination
- In situ structural investigation (currently not feasible)
- Measurement of selected parameters regarding their change due to the initiated friction process (rheometer experiments) to infer grease wear

The latter point can be used as an indirect observation of grease wear.

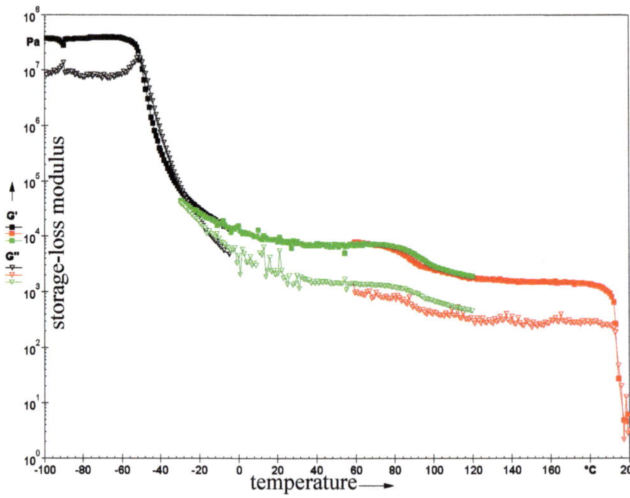

Figure 106: Constant oscillation over a wide temperature ramp

The understanding of the driving force for initiating a wear process can be illustrated by a rheometer experiment over a wide temperature range.

An oscillating shear with an extremely small constant amplitude is applied to a grease sample, and its behavior is observed while thermal energy is supplied within the range of $\vartheta = -100^0C$ to $\vartheta = 200^0C$. Although this is not a typical rheometer experiment for investigating lubricating greases, a universally valid interpretation can still be found. Even after the loss of stability, the system responds through different energy-dissipating mechanisms, finding a path to a new steady non-equilibrium state. This continues until a stable state is (only) achieved through self-destruction (Figure 106). Figure 107 shows the experimental technique used over a wide temperature range to observe emerging instabilities and the transition to stabilities.

In [164], the proposal of a viscosity rate is described (very early) with

$$\dot{\eta} = \frac{\partial \eta}{\partial t} \tag{254}$$

and, using the relationship between frictional energy E_f and wear volume V_V, the expression

$$\dot{\eta} = \frac{\partial e^*_R rheo/\partial t - \tau \cdot \partial\gamma/\partial t}{(\partial\gamma/\partial t)^2 \cdot t} \tag{255}$$

is developed. From this, dependencies of a viscosity rate, which is here regarded as an expression of lubricating grease wear, can be inferred.

One can only speak of the service life, failure, or performance of a lubricating grease in the context of its application. When attempting a detached analysis, the description of

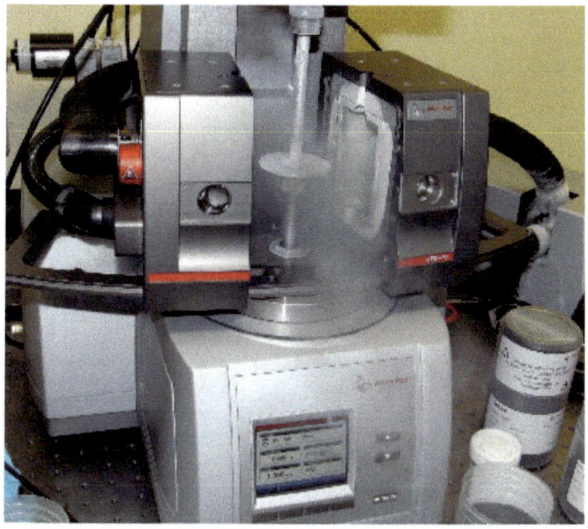

Figure 107: Opening the temperature chamber during the low-temperature experiment on the rheometer [215]

the system's response focuses on the correlation between driving forces and their implementation.

The previously presented studies on structural degradation and the subsequent work on lubricating grease wear aim to create opportunities for comparably representing tribological behavior and quantifying wear. The literature presents very different approaches to this.

In this work, an attempt is made to analyze and describe a tribological system as a result of the input of mechanical energy, using the example of the stressed grease volume. The focus is therefore less on the grease-lubricated contact and more on the stressed lubricating grease.

A testing procedure was developed on the rheometer, which, in addition to the shear test in rotation mode at a constant shear rate, includes an additional section for assessing the process behavior. The entire test consists of three sections:

1. Filling the measurement gap followed by a 15-minute resting phase.
2. Conducting a loading period in rotation mode with a defined shear rate $\dot{\gamma}$ and loading duration t.
3. *Immediately* afterward, an oscillation measurement (amplitude sweep) is performed with a constant frequency f and a deformation ramp. This section is carried out until the crossover point of the storage and loss moduli is reached.

This experimental procedure (see Figure 108) is used for a comparative investigation. The

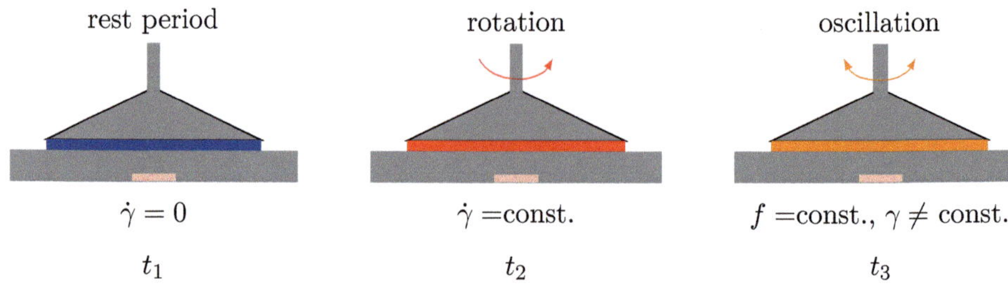

Figure 108: Illustration of the experimental procedure.

mechanical energy E_f expended during the loading section, the energy expended during the oscillation section E_{CP}, and a measurement of the temperature change (during the loading section) are evaluated.

This essentially means that when the structure is comparatively significantly altered during the loading phase, comparatively less energy is required in the oscillation section to reach the crossover point of the moduli, and vice versa.

The experimental temperature ϑ_{shear} and the shear rate $\dot{\gamma}$ in the rotation section can be varied when investigating different model grease samples.

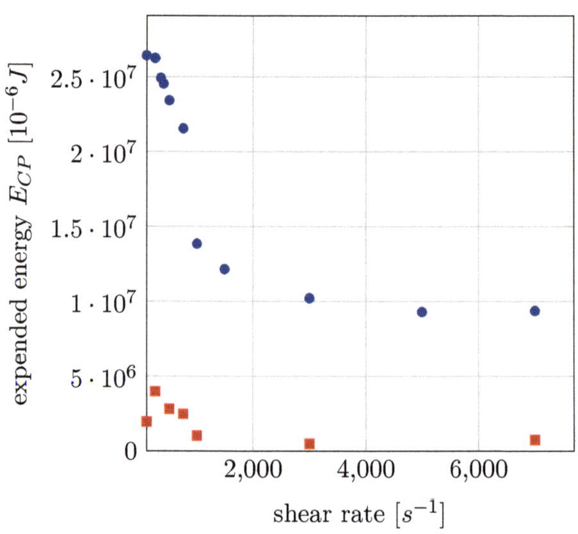

Figure 109: Behavior up to the crossover point for a Li-soap model grease at $\vartheta = 26^0C$ and $\vartheta = 52^0C$

The comparison of the experiment at different temperatures according to Figure (109) shows similar behavior, but it occurs at different levels. At low shear rates during the loading section, the structure is comparatively less altered, and in the subsequent oscillation section, a comparatively high energy expenditure is required. This changes with increasing shear rate in the rotational test. The development of a Li sample and an HD SI sample is shown in Figure (110).

To better compare the wear behavior of grease samples in this experimental procedure, a parameter R_{tee} is introduced [216]. For better understanding, it is used here in the reciprocal form compared to L. AHME ET AL. [216]. It represents the ratio of the mechanical energy expended during the rotational section to the energy expended in the oscillation test and is thus linked to the experiment in Figure (108).

$$R_{tee} = \frac{E_f}{E_{CP}} \tag{256}$$

with E_f as the energy expenditure in the rotational test, E_{CP} as the energy expenditure

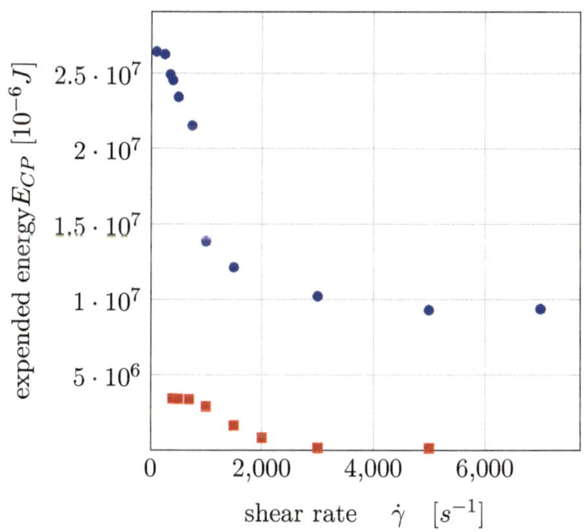

Figure 110: Behavior up to the crossover point for a Li-soap model grease (top) and a Gel model grease (bottom), both NLGI2

in the oscillation test until the crossover point is reached, and R_{tee} as the *wear factor*.

Thus, small R_{tee} values are found when the sample shows only minor structural changes under comparable frictional stress, and vice versa. The *wear factor* varies depending on the sample composition and solid type under otherwise identical experimental conditions.

In Figure (111), three grease samples are exemplarily tested and compared. All tend toward an asymptotic structural state at high shear rates. Under comparable frictional

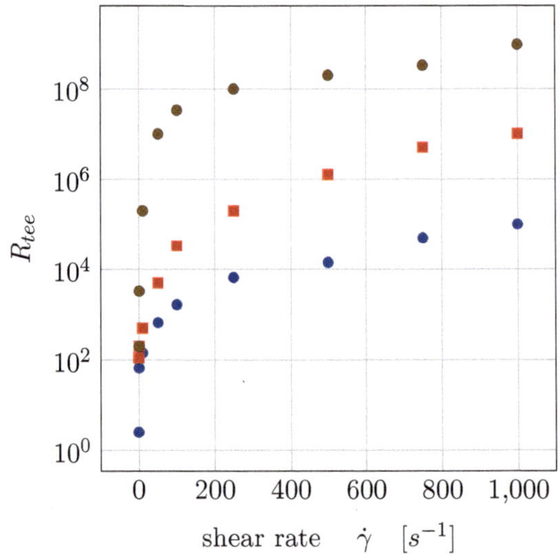

Figure 111: Wear factor for a Li sample, a PU sample, and a biogenic sample according to [216]

stress, the biogenic grease shows the highest wear, while the Li-soap sample shows the least grease wear (see Table (9.1)).

Solid	Solid Content	Base Oil
• Lithium-12-hydroxystearate	9.5%	mineral oil
• Polyurea	19.6%	PAO
• Beeswax, Glycerol monostearate, Cetyl alcohol	7%, 5%, 3%	HOSO, castor oil

Table 9.1: Sample materials for Figure (111)

Temperature Measurements Information on the transformation of the mechanical energy introduced into the grease volume during the friction process is scarce or nonexistent. Therefore, an attempt was made to observe the temperature development during a rheometer test. Initially, the temperature control system of the rheometer was used [13] to record information with the sensor located in the stationary plate under as constant environmental conditions as possible (Figure (112)).

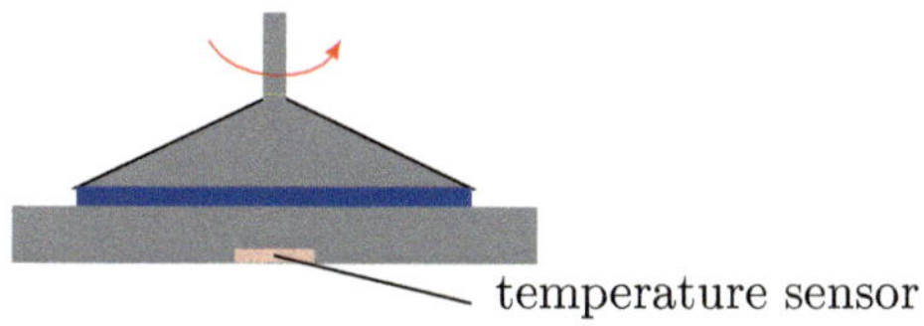

Figure 112: Rheometer geometry and temperature sensor

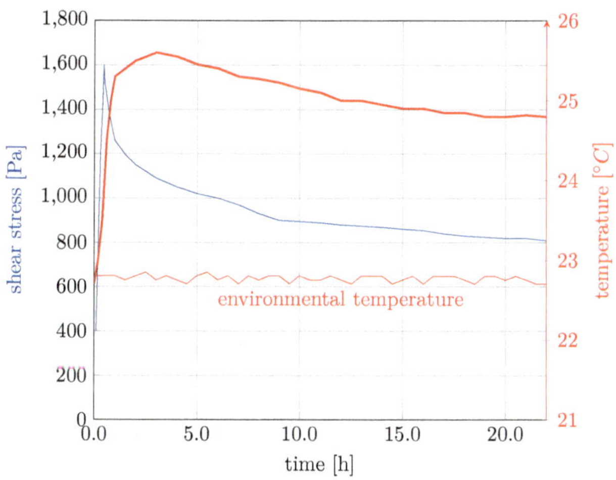

Figure 113: Shear stress, ambient temperature, and temperature development, exemplified by experiments in [217] with $\dot{\gamma} = 500\text{s}^{-1}$

In [217], a series of model greases were observed in terms of heat development. The principal trends of shear stress and temperature are similar in Figure (105) and Figure (113). Below are two model substances: one with Lithium-12-hydroxystearate with 9.5% and mineral oil with $\nu = 240\text{mm}^2\text{s}^{-1}$, and another with 13% PAO base oil with $\nu = 48\text{mm}^2\text{s}^{-1}$, and a sample with Calcium-12-hydroxystearate with 9.7% and castor oil base oil with $\nu = 240\text{mm}^2\text{s}^{-1}$ (Figure (114)). The investigations into the wear behavior and its description using the R_{tee} factor are now correlated with the heat development during the wear process. In [217], the following model greases were used, all belonging to the NLGI2 consistency class (Table 9.2).

The Li- and Ca-based greases in the used sample material exhibit the expected correlation between heat development and wear. The PU grease stands out with extreme wear

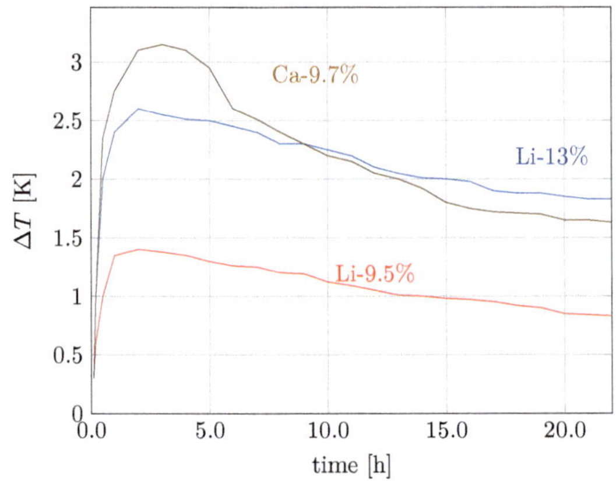

Figure 114: Example of temperature measurement during the rotational test for three model substances according to [217] with $\dot{\gamma} = 500\,s^{-1}$

Sample	Solid	Solid Content	Base Oil	ν_{40} in $[mm^2 s^{-1}]$
C1	Li-12-hydroxystearate	16.1	castor oil	240
C2	Li-12-hydroxystearate	10.6	PAO	240
C3	Li-12-hydroxystearate	13	PAO	48
C4	Li-12-hydroxystearate	9.5	mineral oil	240
C5	Ca-12-hydroxystearate	9.7	castor oil	240
C6	Ca-12-hydroxystearate	22.8	PAO	240
C7	Polyurea	19.6	PAO	240

Table 9.2: Sample materials for Figure (115)

but similar heat development. It should be noted that the second y-axis is logarithmically scaled. The wear behavior was investigated at a test temperature of $\vartheta = 40°C$. The biogenic sample $B3$ from Table (9.1) reacts quite differently.

For further investigations into heat development and distribution, the *Calidus* system was developed. It includes several sensors directly on the surface of the lower rheometer plate (Figure (116)). As expected, the measured temperatures are higher than those from the standard rheometer sensor. The system is still under development.

9.3.4 Activation Energy and Grease Wear

An initiated friction process in the grease film inevitably leads to a change in the grease structure, known as *grease wear*. This process involves the input of mechanical energy,

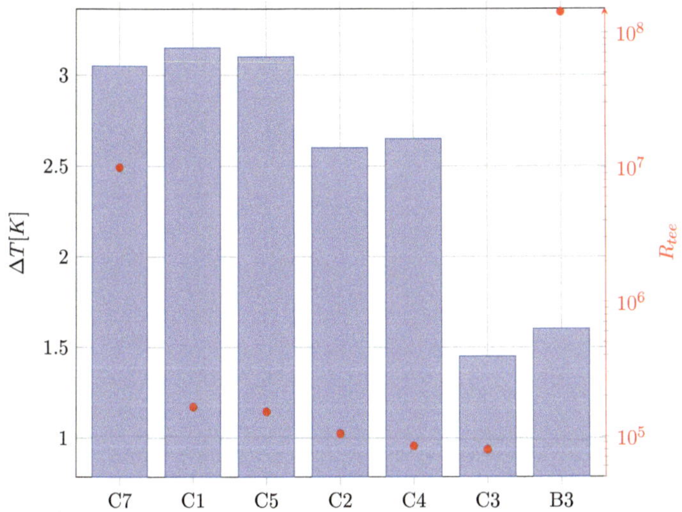

Figure 115: Temperature development and wear factor for the samples listed in Table (9.2).

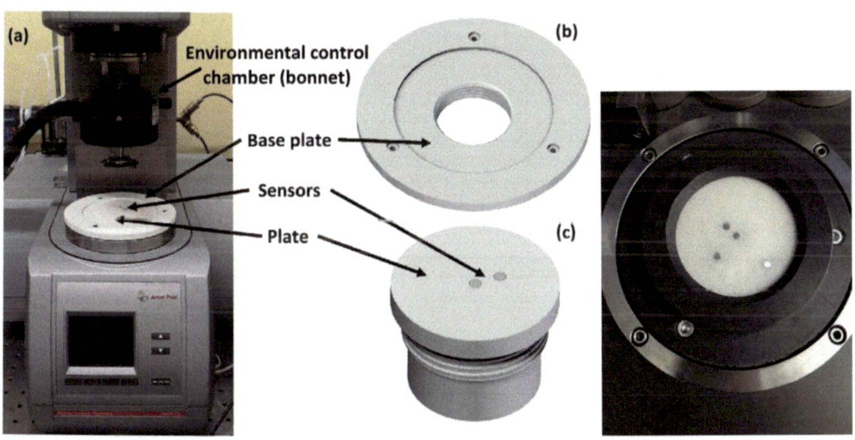

Figure 116: The Calidus measurement system [217], with left showing its positioning in the rheometer and right showing the arrangement of multiple sensors

which activates a structural change and thus grease wear.

The activation energy E_a describes the amount of energy required within a certain temperature range that molecules need to trigger the wear process [72]. Using the energy densities $e_{rheo}[Jm^{-3}]$ from a rheometer rotational test (Equation (242)) and its temperature dependence, we have

$$e_{rheo} = A \cdot e^{\frac{E_a}{R \cdot T}} \tag{257}$$

or

$$\ln e_{rheo} = \ln A + \frac{E_a}{R \cdot T} \tag{258}$$

where $A\,[Jm^{-3}]$ is a pre-exponential factor, $E_a\,[Jmol^{-1}]$ is the activation energy, $T\,[K]$ is the absolute temperature, and $R\,[Jmol^{-1}K^{-1}]$ is the gas constant. At infinitely high temperatures, A represents the limiting energy density [72].

$$e_{rheo}(T \to \infty) = A \tag{259}$$

The ratio E_a/R can be determined experimentally. The logarithm of e_{rheo} is plotted against the reciprocal value of T. N. Acar [122] shows that for the Arrhenius plot, the

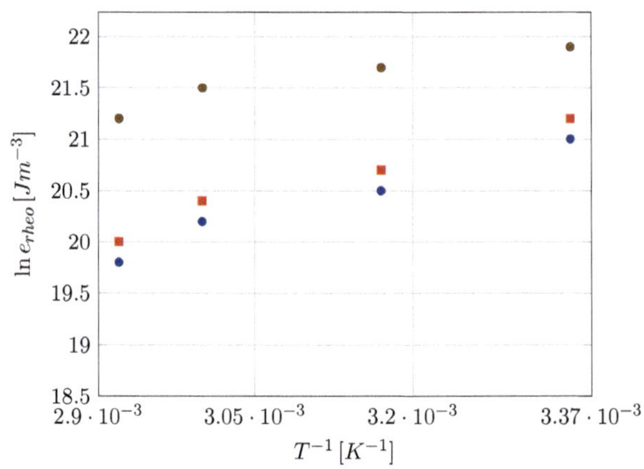

Figure 117: Example of an Arrhenius plot for three grease samples from NLGI classes 0 (bottom), 1, and 2 (top).

activation energy E_a can be expressed as

$$E_a = R \cdot \frac{\ln e_T}{(T^{-1}) - (T_{ref}^{-1})} \tag{260}$$

where

$$e_T = \frac{e_{rheo(T)}}{e_{rheo(T_{ref})}} \tag{261}$$

and thus, the energy density at T_3 is given by

$$e_{rheo(T_3)} = e_{rheo(T_{ref})} \cdot e_T \tag{262}$$

Using amplitude sweeps in oscillation tests, it can be observed that with increasing solid content, the critical deformation decreases. This is due to the weak physical bonds in the agglomerate and the strong chemical bonds in the base oil [150]. When correlating the activation energy with the critical deformations from the oscillation tests, Figure (118) illustrates this relationship.

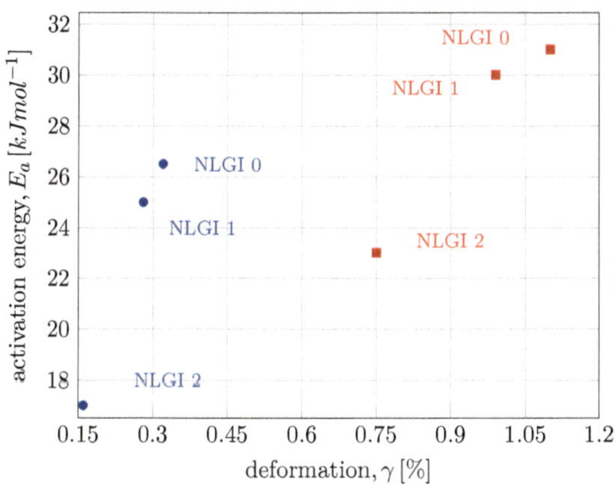

Figure 118: Activation energy and deformation. On the left: Li-based model greases, and on the right: PU-based model greases according to [72].

A similar effect is observed when parameters other than the critical deformation γ are investigated. When examining the change in the storage modulus G' during an amplitude sweep, between the onset of plastic deformation and the crossover point, the behavior is as shown in Figure (119).

9.3.5 Remarks on EHL with the Presence of Grease

At the end of the 19th century, H. Hertz published a paper on the contact of elastic solid bodies [218]. Almost simultaneously, B. Tower [219] discovered pressure build-up in a lubricant film during bearing tests. Since then, there has been an ongoing development in both the theoretical and experimental description of lubricant film thickness and pressure build-up in heavily loaded contacts. This development has involved many milestones such as O. Reynolds [220] and A. Mohrenstein-Ertel [221], as well as numerous other notable names (see also [222], [223]).

In the context of rolling bearing lubrication, there are numerous studies on grease-lubricated friction contacts. Such highly stressed friction pairs generally operate under

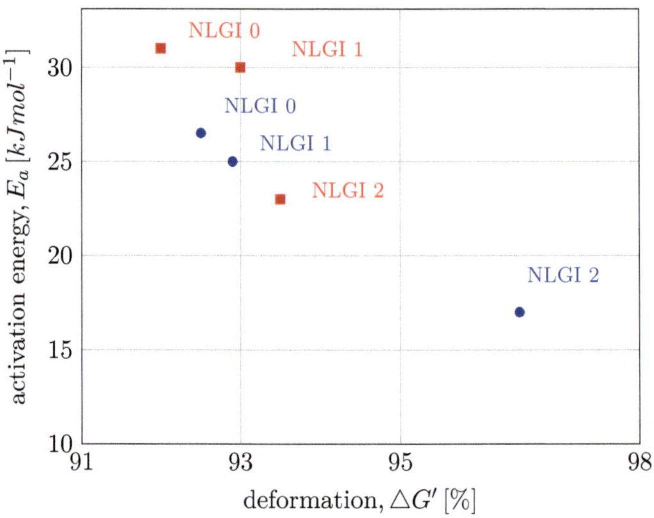

Figure 119: Activation energy and change in storage modulus: Li-based model greases and PU-based model greases according to [72]

what are known as Elasto-Hydrodynamic-Lubrication (EHL) conditions, and their study is also of interest for fundamental tribological questions. These conditions imply the presence of two friction states: solid friction due to elastic deformation of the solid bodies, and fluid friction due to the stress on the lubricant.

For the description of gap conditions and lubricant film thickness measurements in ball-on-flat contacts, there are many fundamental works. Interest in this topic increased in the 1990s and 2000s due to rapid advancements in experimental techniques for measuring lubricant film thickness, e.g., [224], [225], [226], [227], [228], [229].

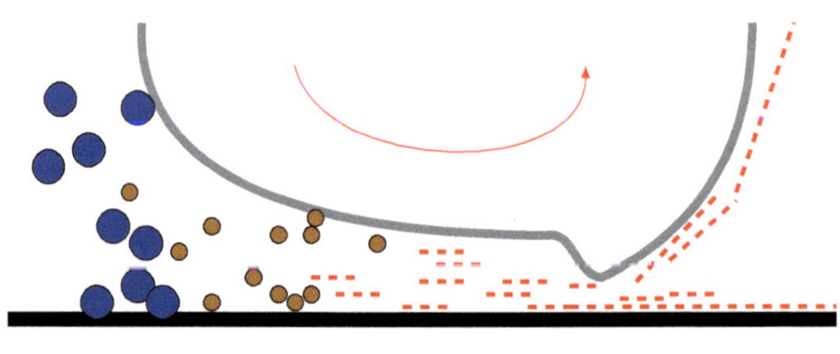

Figure 120: General overview of grease wear in EHD contacts

The general understanding of structural changes in the grease film has been developed since the 1990s and is illustrated in Figure (120). The macroscopic agglomerates are broken down into smaller structures by the friction process, and in the highly stressed areas, they are eventually transformed into micro- and nano-elements. Upon exiting the gap, behind the Petrusevich peak, a concentrated layer remains on both contact surfaces. At the exit of the gap, in a ball-on-disc contact, tensile stresses on the lubricant are conceivable.

The peculiarities of grease behavior in the highly loaded gap result from its composition, structural build-up, and distinctly viscoelastic, rheological behavior. Experimentally, investigations into lubricant film formation are carried out using a ball-glass disc arrangement and interference measurements.

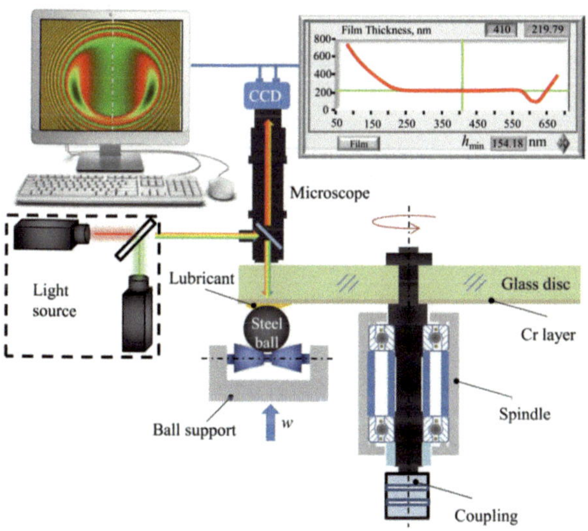

Figure 121: Schematic diagram of the test rig and measurement scheme from [230]

This technique has rapidly developed and provides a wealth of information about the stressed lubricant film. When applying this investigative method, it is important to distinguish between a fully flooded and a starvation contact. To establish a fully flooded condition in the experiment with the setup shown in Figure (121), Fischer et al. [231] used the model depicted in Figure (122).

For the starvation condition, they employed a scraper as shown in Figure (123).

The fully flooded contact shows the typical EHD lubricant film thickness profile, whereas the contact that is only sufficiently lubricated forms the Petrusevich peak to a lesser extent. Investigations into this have been conducted, among others, by Gonçalves et al. [232] and are shown in Figure 124.

Profile lines parallel and perpendicular to the direction of motion illustrate the geometric

Figure 122: Device ensuring fully flooded contact [231]

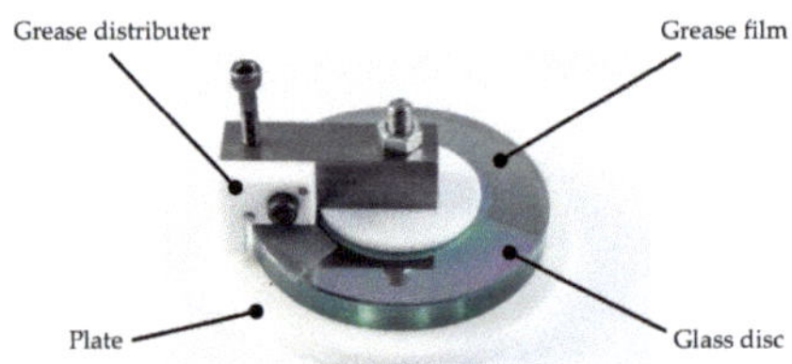

Figure 123: Grease distributor for starvation condition studies

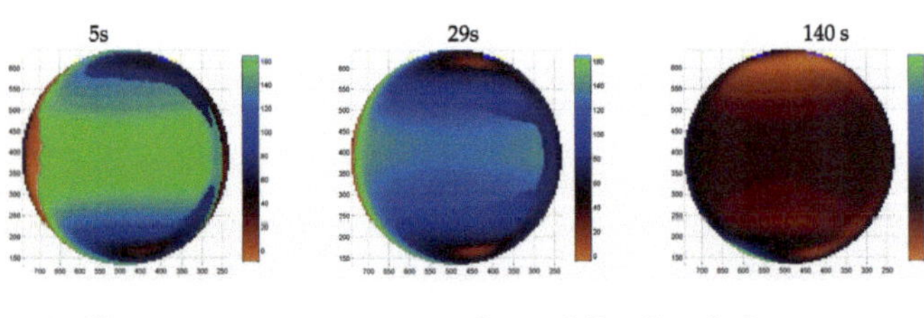

Figure 124: Change in gap geometry from fully flooded to starvation [232]

relationships and their changes.

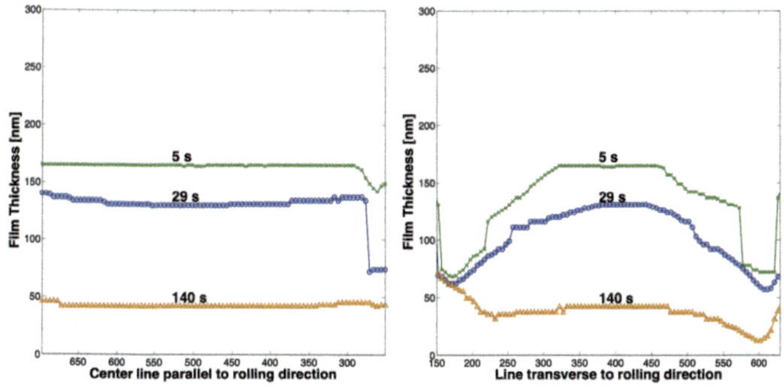

Figure 125: Cross sections parallel to the rolling direction (bottom left) and transverse to the rolling direction (bottom right) [232] related to Figure (124).

The development of the gap conditions over time is further illustrated in a different representation by GONÇALVES ET AL.

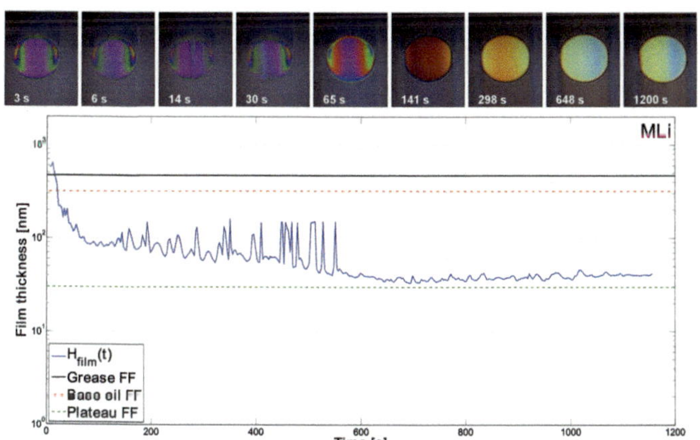

Figure 126: Central film thickness of a ball-on-disc contact, U = 0.5 m/s, F = 50 N, SRR = 5%, and T = 40°C, [232]

It is interesting to note the more or less stable level of lubricant film thickness that develops, which can only be disturbed by larger agglomerates. In the diagram, (FF) denotes fully flooded.

GONÇALVES, CAMPO and SEABRA[232] reached the following conclusions:

- An increase in relative speed leads to quicker establishment of sufficient lubrication.

- An increase in temperature slows down the formation of a starvation contact due to better lubrication supply to the contact pair.

- A rising slide-to-roll ratio accelerates the transition from fully flooded to starvation. Simultaneously, the film thickness decreases due to the shearing process.

The comparison of base oil, fully flooded grease contact, and grease starvation is shown by [231] in Figure (127).

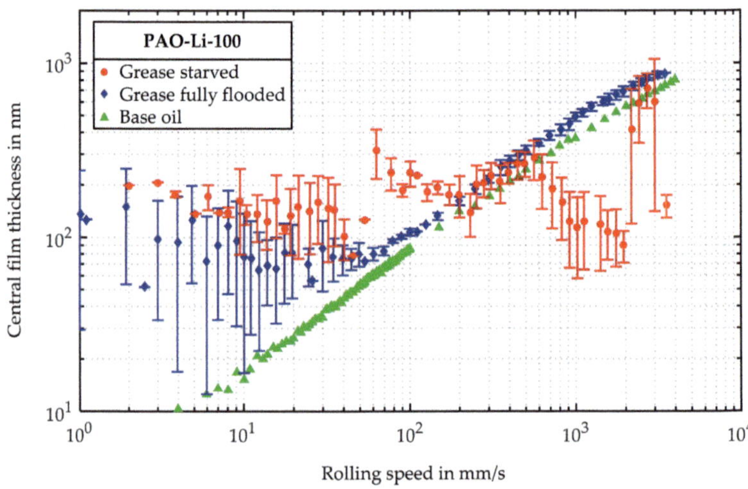

Figure 127: Film thickness of base oil, grease under fully flooded, and starved lubrication for PAO-Li-140 [231].

At a rolling speed of approximately 600 mm/s, starvation sets in. Until that point, the lubricant film thickness is greater than that of the base oil and the fully flooded contact.

Li et al. [230] investigate the influence of coarse and fine fiber solid thickeners on EHD lubricant film formation. The samples they used are shown in Figure (128).

Their investigations yield the following insights:

- "At low speeds, both types of grease, with thin and coarse fibers, form a thicker film than the corresponding base oil, which is attributed to the transfer of thickener lumps. The grease with coarse fibers forms large lumps that cause significant fluctuations in film thickness.

- At moderate speeds, there is a progressive film breakdown due to starvation.

- At higher speeds, the grease with thin fibers is completely sheared and degraded, resulting in a highly viscous lubricant that can generate a thick film. In contrast, the

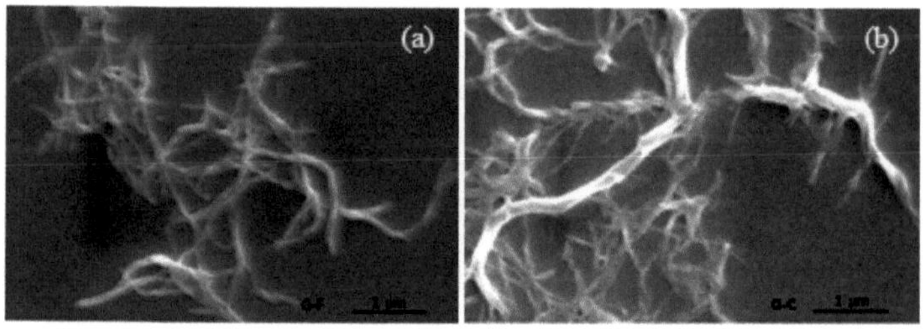

Figure 128: Structure of the model greases used by [230]

grease with coarse fibers exhibits a high bleeding rate, leading to increased lubricant loss and more pronounced film degradation.

- Both the residual layers and the hydrodynamic films contribute to film formation in grease lubrication. The grease flow and distribution on the raceway sides affect lubricant replenishment and the formation of equilibrium films."

9.3.6 Other Experimental Investigations on Grease Behavior

Acoustic Measurements To obtain additional, different types of information on grease wear, I initiated experiments in 2009 to measure acoustic emissions during the degradation of grease structure [233], [234], [235].

The acoustic emission measurements were conducted using a Physica Rheometer MCR 300. A plate-plate system was selected to study a total of nine grease samples with biological base oils and one reference base oil. The sample with sunflower oil as the base oil proved to be particularly suitable for the measurement system used. Preliminary investigations determined that the optimal position for the acoustic emission sensor was at the center of the lower rheometer plate. No coupling medium was required, as the sample under investigation served as both the coupling medium and the test object. A measurement system from *Physical Acoustics* was used, with the settings listed in Table 9.3. An amplitude threshold was set to suppress noise and record only signals from the stressed sample structure. The lower plate was prepared for sensor installation, as shown in Figure (129).

In the oscillation test (amplitude sweep), the intersection point of both modules is considered a significant marker. This point is viewed as the achievement of a state where the entire sample volume is structurally altered. The signals are described by the maximum peak in dB_{AE} and the absolute energy in aJ.

For selected model greases, favorable test conditions are determined in preliminary experiments. The aim is to identify emission differences in samples with varying compositions. The samples are subjected to oscillatory stress beyond the linear-viscoelastic region (LVE) and are simultaneously observed with the acoustic emission sensor.

Instrument	PCI-2 2-channel system, Physical Acoustics
Preamplifier	2/4/6 Preamplifier, Physical Acoustics
Sensor	PAC WD, 100 kHz - 1000 kHz, Physical Acoustics
Preamplifier Gain	40 dB_{AE}
Preamplifier Filter	100 kHz - 1200 kHz
Threshold	24 dB_{AE}
Sampling Rate	10 MHz
Measurement Variables	Amplitude in dB_{AE} and absolute energy in aJ

Table 9.3: Settings used for the acoustic emission measurement

Figure 129: Installation of the acoustic sensor [233]

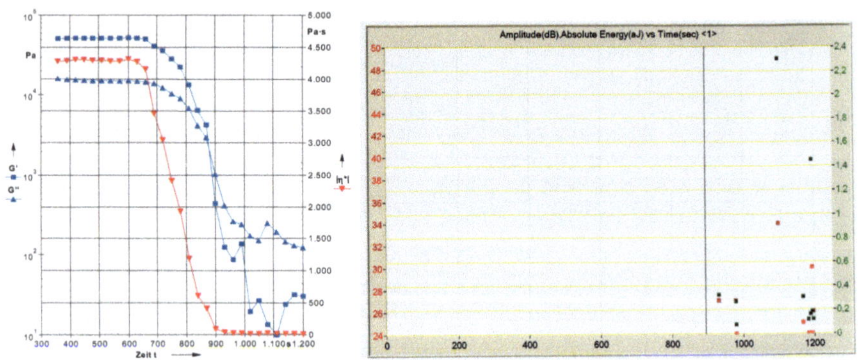

Figure 130: Amplitude sweep and parallel acoustic emission measurement [233]. The marker in the right diagram indicates the position of the intersection point.

Typically, the observations appear as shown in Figure (130). *Hits*, i.e., emissions, were recorded almost exclusively after reaching the intersection point. To conduct measurements with different sample materials, seven biogenic grease samples from *Universidad de Huelva* were selected, and amplitude sweep tests with acoustic emission measurements were performed. All samples were made with castor oil as the base oil, as detailed in Table 9.4.

1	ethyl cellulose, alpha cellulose -1
2	ethyl cellulose, alpha cellulose -2
3	ethyl cellulose, alpha cellulose -3
4	chitin
5	chitosan
6	sorbitan monostearate
7	alpha cellulose

Table 9.4: Bio-additives of the investigated biogenic samples

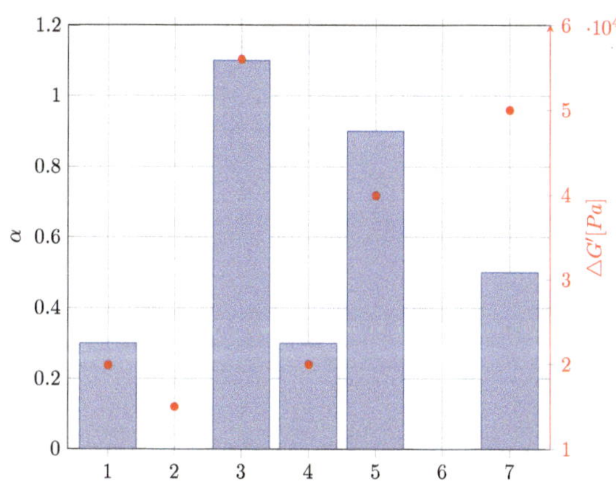

Figure 131: α-values of the investigated samples and the corresponding change in the storage modulus during amplitude sweep

The analysis includes, among other things, the sum of the absolute energy relative to the number of hits, i.e., the energy per hit $= \alpha$, as well as the maximum amplitude in dB [235], the rheological parameters, and, for example, the change in the storage modulus $\Delta G'$. Figure (131) illustrates a suspected correlation between energy per hit and the absolute change in the storage modulus.

It appears that intense structural degradation, indicated by a significant decrease in the G' value, leads to high energy emissions. Samples 2 and 6 showed no measurable emission. One could also interpret that sample materials with a well-developed solid structure, which can store energy through elastic deformation, emit relatively high-energy hits during the transition to the plastic region and intense structural degradation.

In a further experiment, continuous loading for 7 h and a deformation of $\gamma = 1000\%$ were observed with the acoustic emission sensor. The emission activities were expected to exhibit a temporal behavior similar to the τ-profile of a rotational measurement.

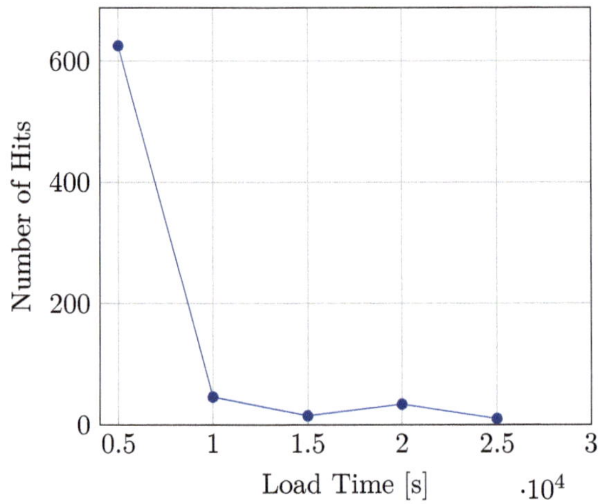

Figure 132: Number of hits in the 7h continuous test [233]

The number of hits is always considered within ranges up to 5000 s, from 5000 s to 10000 s, and so on (Figure (132)). As expected, the process is concentrated in the initial period and then tends to approach a quasi-steady state.

A test procedure with a pause as a rest period also reveals a process behavior that is well-known from other experimental setups. In the initial loading period, a high signal density with relatively high-energy hits is observed. The measurement system seems unsuitable for recording activities during the subsequent rest phase. The second loading period shows a more intense emission at the beginning compared to the end of the first phase. This then falls to a steady level as usual.

It is suspected that the initial intense structural degradation moves towards a steady state, followed by a rest period with weak structural regeneration, and then a repetition of the loading process (Figure (133)). Further experiments on thixotropic behavior could be of interest.

Ball-Drop Experiments A phenomenological study of the impact loading of a grease volume was initiated with the so-called ball-drop experiments. These have been documented in [236], [111], and [237], among others. The objective was to investigate the varying impact on a steel sample (flat) when a steel ball impacts a defined layer of grease on the surface of this steel plate. Additionally, the grease filament lengths produced during the ball rebound were analyzed (video).

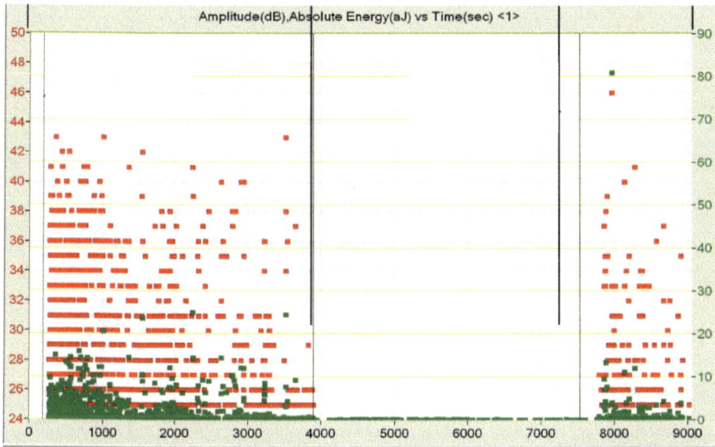

Figure 133: Loading-Rest-Loading cycle with acoustic emission measurement [233]

A 1mm thick layer of the sample material was applied to a ground stainless steel plate. The ball had a diameter of 25mm and was dropped from a height of 1m. For each grease sample, 18 repetitions were conducted. New biogenic grease samples were used for the experimental investigations at that time.

Initially, it was not the measured calotte diameters and calotte depths that were of interest, but rather the appearance of the calotte surfaces. It was expected that during the experiment, ball impressions with varying diameters would be observed, an assumption that can be justified by the different damping behaviors. However, what was not anticipated was the occasionally striking structuring of the calotte surfaces (Figures (134) and (135)).

A commercial grease with synthetic ester and Li-/Ca-soap was used as a reference lubricant (Figure (136)). Trials without lubricant were also conducted for evaluation.

In the evaluation of calotte depths, two model substances were found to have depths lower than that of the reference grease (Figure 137). These were a model grease with sunflower oil and castor oil, and another with glycerin, cetyl alcohol, and sorbitan fatty acid esters. The second sample consisted of sunflower oil and castor oil with ethyl cellulose and carnauba wax. The mean depths were $35.7\mu m$ for the reference grease, $35.5\mu m$ for the sorbitan fatty acid esters, and $34.9\mu m$ for the carnauba wax. The mean depth without lubricant was $37.2\mu m$.

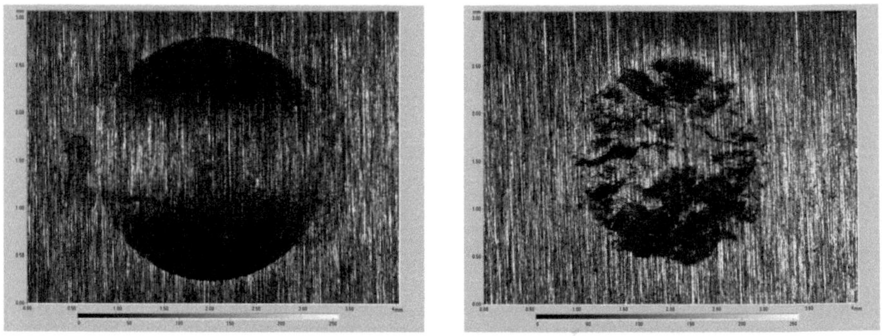

Figure 134: *Left:* Sunflower oil, glycerin, and cellulose; *Right:* Sunflower oil and corn cob meal

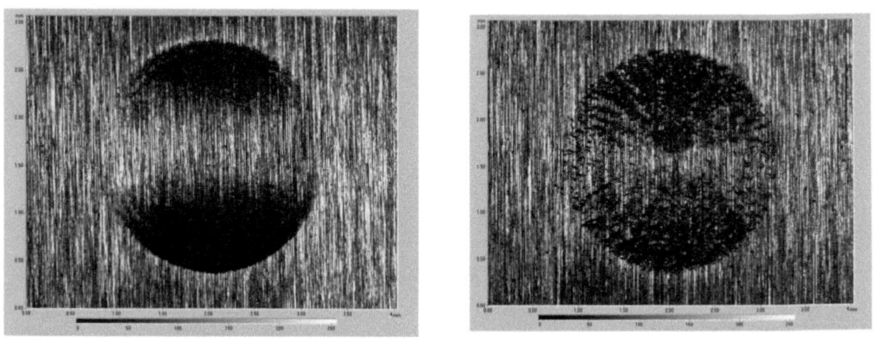

Figure 135: *Left:* Castor oil, sunflower oil, and butyric acid; *Right:* Sunflower oil and lignosulfonate

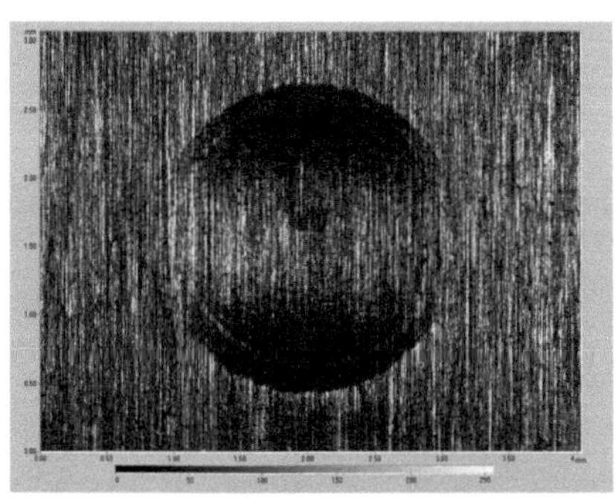

Figure 136: Calotte of the reference grease

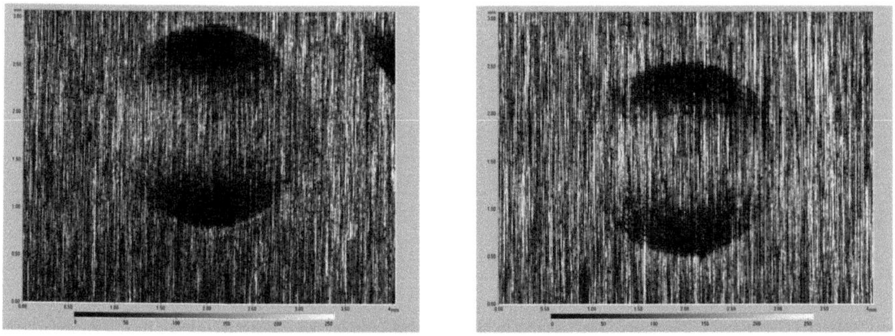

Figure 137: Ball calottes of the model greases with the smallest calotte depths

Some Aspects of the Influence of Oil Polarity To investigate the influence of base oil polarity on various aspects of the tribological process, the focus was placed on Keesom forces, which are the strongest secondary valence bonding forces. To experimentally analyze their influence, experimental conditions were chosen that do not counteract the formation of Keesom forces. The strong temperature dependence of these bonding forces is responsible for selecting the temperature range of the experiments, $\vartheta = -10°C$ to $80°C$. The impact on friction in the grease film can be represented by examining the energy density applied during a shear experiment in a rheometer. The high-polarity base oils used were high oleic sunflower oil (HOSO) and trimethylolpropane trioleate (TMPO). Low-polarity oils included polyalpha olefin (PAO) and octyldodecyl isostearate (OCT). Figure 138 shows that at the experimental temperature of $\vartheta = -10°C$, the influence of polarity is most pronounced. At higher temperatures, the difference disappears as physical bonds become weaker. Results with additional model greases can be found in [111].

The wear behavior of a steel-steel pair in a ball-on-disc contact (sliding friction) is illustrated in Figure 139. Here, the positive effect of the base oil polarity is particularly evident. Shown is the solid body wear on the ball. The combination of sapphire-steel was also investigated in [110], with both ball and disc wear analyzed. Considering the activation energy as shown in [238], there is also an observable influence of the base oil polarity (Figure 140).

It appears that for highly polar base oils, higher activation energies are required to induce structural degradation under transient shear. Detailed results and further findings are presented in [238].

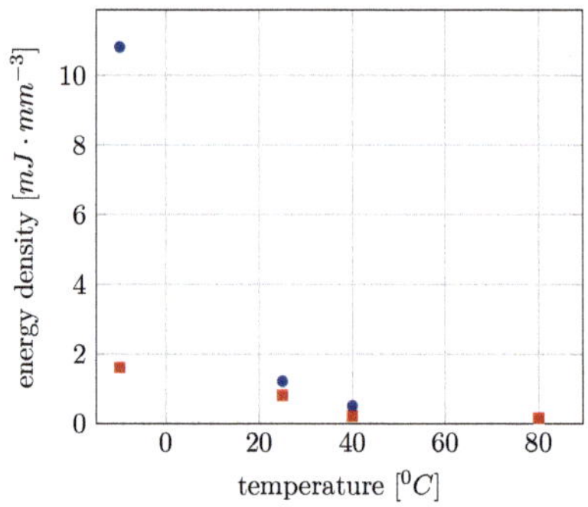

Figure 138: Energy density over the test temperature, high polarity (HOSO) - low polarity (OCT)

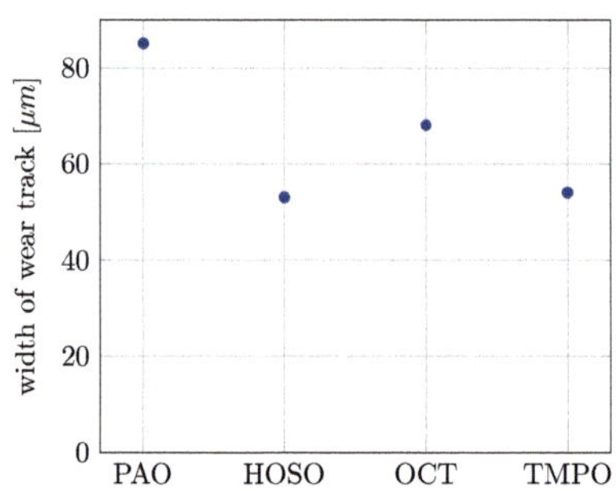

Figure 139: Width of the wear track on the ball

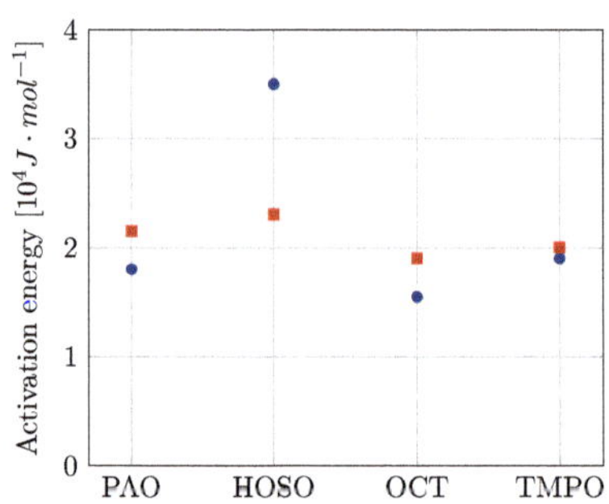

Figure 140: Activation energy in transient shear flow from [238], with highly dispersed silica acid and lithium solid (thickener)

9.3.7 On Thixotropy

For this section, I have adopted the heading chosen by HERBERT FREUNDLICH in his 1928 paper, not only to acknowledge his extremely important work but also to remember an extraordinary individual with a tragic fate (see Figure (141)) and also the information provided in Figure (57)). From the FREUNDLICH paper [143], the following quote is

Figure 141: Prof. Herbert Freundlich, source Wikipedia

provided:

"I would first like to show you the appearance in the form in which it was first discovered by SZEGVARI and FRL. SCHALEK [141], [142]. Here you see a concentrated, approximately five percent iron oxide sol - the well-known solution of Ferrum oxydatum, dialysatum liquidum - that was made to solidify into a paste-like gel by adding a suitable amount of salt. By shaking, it liquefies into a sol and solidifies, when left to itself, into a gel in a clearly defined time, which depends on temperature and other conditions. This process can be repeated as often as desired; the solidification time remains constant under the same external conditions. This isothermal reversible sol-gel transformation is usually referred to by us, following a suggestion by PETERFI, as *thixotropy*. Such a gel is accordingly termed *thixotropic*."

FREUNDLICH then undertakes a variety of very different experimental and theoretical investigations that shape the representation and understanding of this phenomenon [145], [239], [240], [241], [242].

For the purpose of this book, it is noteworthy that in [243], FREUNDLICH and RAWITZER already developed, among other things, an experimental procedure in 1927 that resembles today's thixotropy experiments for studying viscoelastic lubricants.

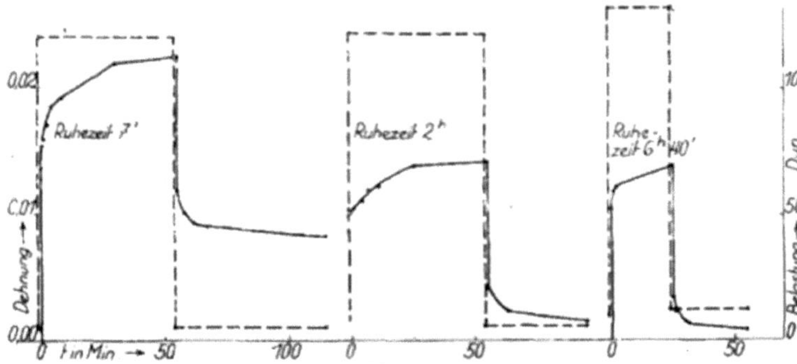

Figure 142: Dashed lines: Load, solid lines: Deformation from [243]

From the experiments in [243], it is concluded:

1. Considerations according to which progressive deformation can be viewed as a pure flow of the sol, such that after infinitely long loading, it has completely yielded and no finite resistance remains, do not apply to the present structures.

2. When a sol is loaded or unloaded, it strives, as long as the breaking point is not exceeded, towards an equilibrium state at first with high speed and then with decreasing speed, which depends on the initial state of the sol.

FREUNDLICH'S experiments and analyses were groundbreaking for understanding *thixotropic* behavior. His observations regarding the achievement of equilibrium states can be re-evaluated from today's perspective of energy balance.

9.3.8 On Lubricating Grease Thixotropy

The investigations into the process phenomena under shear stress and the removal of this stress differ significantly from the experimental work of FREUNDLICH. The experimental procedures and observed effects are quite different from the original studies with iron oxide sol. To distinguish this, the term *grease thixotropy* is proposed, which highlights the differences from the original work and its original meaning in this word combination.

In the research on the observed special behavior of lubricating greases after load removal, it is R. CZARNY who has pioneered with his systematic approach [84], [189], [66] and see Figure (143).

He applies the phenomena studied by FREUNDLICH to the specific characteristics of lubricating grease and develops an experimental procedure *stress - resting time - stress*, using a rheometer. Regarding the behavior of lubricating grease, he explains in [189] that a structural change also affects the rheological parameters, including viscosity and shear

Figure 143: Prof. Dr. Ryszard Czarny, a pioneer of investigations into grease thixotropy

stress. An idealized representation of the experiment according to CZARNY is shown in Figure (144). The processes idealized in the course shown in Figure (144) are described by

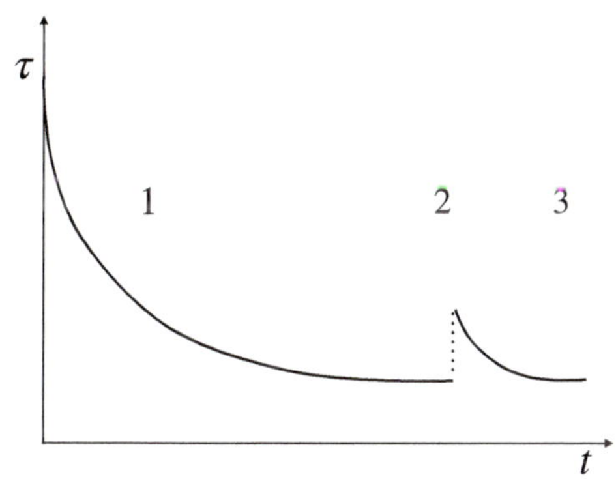

Figure 144: Rheometer experiment on lubricating grease thixotropy: 1- shear stress, 2- resting time, 3- shear stress

CZARNY as follows: "The structural changes in the lubricating grease are actually caused by two opposing processes. On the one hand, the applied shear force acts to break the structural bonds (see also Equation (192)). Simultaneously, however, the contact of the broken and displaced structural elements promotes new bonds, thus partially rebuilding the structure during the shearing of the grease [84]."And in the same work: "The dynamic equilibrium state is eventually reached when the rate of bond breakdown is balanced by the rate of bond reformation."

Further work was then carried out by M. PASZKOWSKI and S. OLSZTYNSKA-JANUS

[208], [209]. They describe in detail the effects during the resting phase and present measurement results. In doing so, they develop an understanding of the structural behavior during the stress phase and the resting period.

Stress Phase

1. The pattern of the microstructure of fresh unsheared lithium grease. The intact thickener microstructure forms a space structure in which soap floccules are long and strongly cross-linked.
2. The first minutes of shear. The cross-linked microstructure largely undergoes disintegration while the more strongly cross-linked aggregates undergo elastic and plastic deformation. Individual fibers freely suspended in oil get oriented during the flow.
3. The microstructure of the thickener after 10 minutes of flow. Visible orientation of thickener fibers and formation of long chains as a result of the formation of additional hydrogen bonds between carbonyl and hydroxyl groups. There is a marked decrease in hydrogen bonds between hydroxyl groups at C_{12}, resulting in fewer soap particle bondings.

Rest period

1. The thickener microstructure immediately after shearing; the floccules are still oriented. But the long thickener chains are unstable and quickly undergo disintegration.
2. The first minutes of relaxation. As the shearing force decays, the orientation of the floccules changes. In this case, the mobility of the soap particles is mainly determined by the viscosity of the dispersion medium.
3. The thickener floccules after $\approx$ 10 minutes of relaxation. The microstructure gets partially reconstructed and the sides of soap particles get stuck together. The formed aggregates have a large number of *free ends* due to an increase in the number of hydrogen bonds between hydroxyl groups.

As an important result of the studies [208], it was found: „The studies have shown that the hydrogen bonds destroyed during shearing reform. Additionally, it was found that the number of COOH groups decreases, indicating a change in the chain structure of the floccules. The greatest increase in the structural viscosity of the grease after shearing and a 24-hour relaxation occurs at low shear rates during the first shear stage. In such cases, the resulting microstructure is the most durable. This is confirmed by the restoration times of the microstructure, determined from the flow curves obtained from the rheometric tests.“

In the work [209], the authors develop a *Three Interval Thixotropy Test (3ITT)* on the rheometer. This means that a lubricating grease sample is subjected to a deformation of $\gamma = 0.1\%$ and a frequency of $f = 1Hz$ in the linear viscoelastic region. First for $t = 100s$ in the first section, and later in the third section for $t = 24h$. In the second interval, $t = 1h$, and the shearing is performed with $\dot{\gamma} = 8.1s^{-1}$. For further analysis, they conduct ATR-FTIR spectroscopy and show, among other things, that

- the investigations indicate a reduction of $COOH$ groups interacting with $-OH$ groups,
- as a result, $-OH$ groups are released by the action of shear forces and can interact with other parts of the Li-12-hydroxystearate molecule chain [209]. This can lead to a structural change in the lubricating grease.

The experimental procedure according to $3ITT$ is similar to the descriptions in Figure (101).

9.3.9 Own Investigations on Lubricating Grease Thixotropy

Experimental investigations

To explore the question of how a lubricating grease sample reacts after being subjected to frictional stress and what the driving forces might be, rheometer tests were conducted, informed by the findings of FREUNDLICH and the experimental work of PASZKOWSKI and OLSTYNSLA-JANUS. These tests also consisted of 3 experimental sections, but with a slightly different focus, as shown in Figure (145). In the first stress section, a sample is

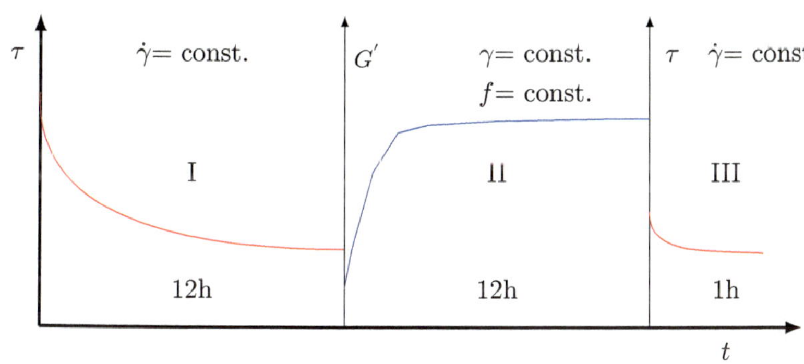

Figure 145: Procedure of the rheometer investigation on lubricating grease thixotropy with I: stress phase, II: resting phase, III: stress phase [244]

sheared at a constant shear rate ($\dot{\gamma} = 1000s^{-1}$ or $\dot{\gamma} = 3000s^{-1}$) and a constant temperature ($\vartheta = 35°C$ or $\vartheta = 50°C$). During the resting section, the temperature is maintained, and the sample is subjected to oscillating stress with an amplitude of $\gamma = 0.02\%$ at a frequency of $f = 1Hz$ in the linear viscoelastic range. After that, the sample is stressed again in rotational mode with $\dot{\gamma}$=const. from the first section.

Initially, experiments were conducted with a Li-based model grease of the NLGI 2 grade at different shear rates during the stress phase (Figure 146). As expected, the still intact structure rebuilds more intensively at a higher level. A similar effect can be observed when considering samples with different solid content (Figure 147). The NLGI 2 sample

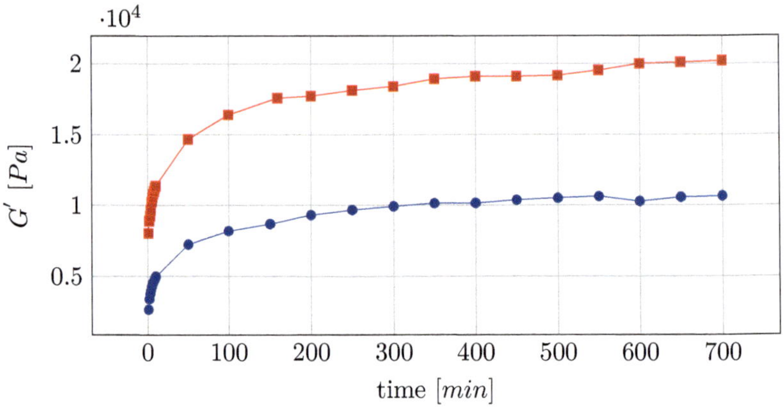

Figure 146: Storage modulus during the rest period II for $\dot{\gamma} = 1000 s^{-1}$ and $\dot{\gamma} = 3000 s^{-1}$

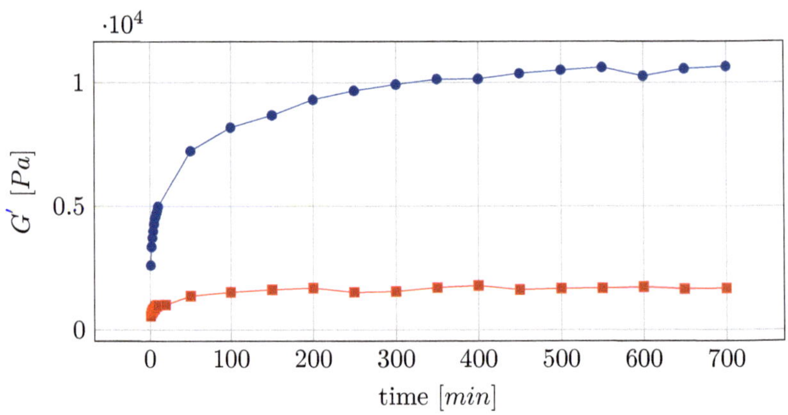

Figure 147: Storage modulus during the rest period II for NLGI 2 und NLGI 1, $\dot{\gamma} = 3000 s^{-1}$

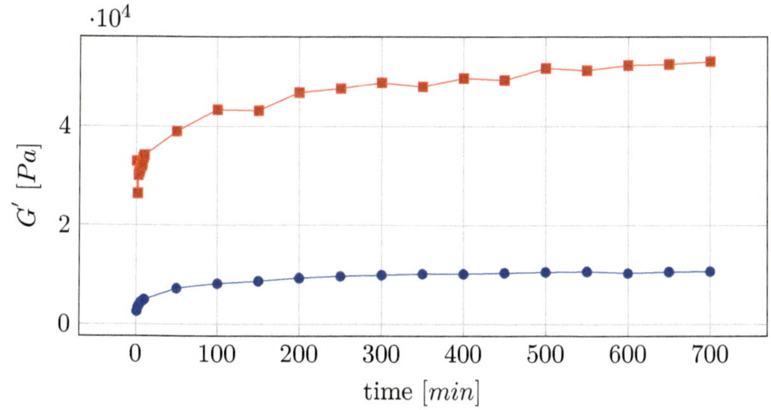

Figure 148: Storage modulus during the rest period II for Li-NLGI 1 und PU-NLGI 1, $\dot{\gamma} = 1000 s^{-1}$

contains 12.9% Li-hydroxystearate as a solid, while the NLGI 1 sample contains 9.8%. In Figure (148), two different solid types were compared: a Li soap (Li) and a diurea (PU), both of which respond very differently to the stress but result in a structural buildup that in both samples tends toward a steady state. When the experiment is conducted at different temperatures (Figure 149), the structural buildup is similar.

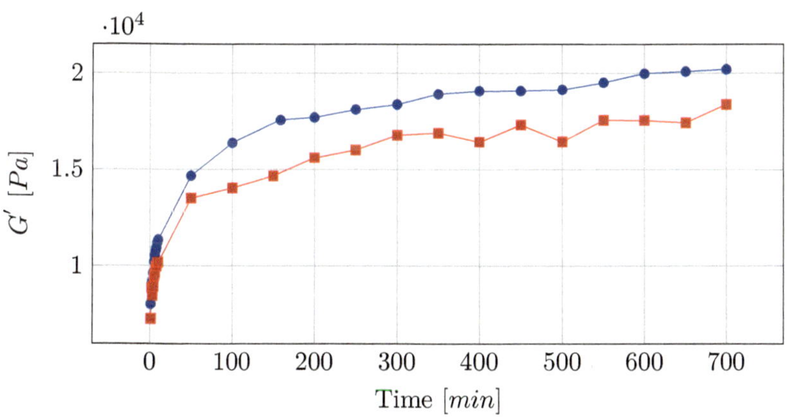

Figure 149: Storage modulus during the rest period II for $\dot{\gamma} = 1000 s^{-1}$, $\vartheta = 50°C$ and $\dot{\gamma} = 1000 s^{-1}$, $\vartheta = 35°C$

Remarks on the Driving Forces of Thixotropic Effects

In the majority of investigations into the thixotropic effects of stressed lubricating greases, the focus is on the difference between the measured parameters at the end of a loading phase and the starting value after a longer rest phase [189], [208]. Considerations are now being made to explore the driving forces behind the mechanisms [209] occurring during the rest phase .

This should direct our attention to different levels: the *macro level*, the *meso (micro) level*, and the *nano level* where we can make different observations.

⟶ *Macro level*: Continuum material characterized by macroscopic properties describable with storage modulus G', yield stress τ_Y, apparent viscosity η^*, etc. These are correlated with the respective structural state.

⟶ *Meso level*: Determined by the structural shape, i.e., the arrangement and geometry of the structural elements, individual fibrils, agglomerates, changes in the degree of order through structural reconfiguration.

⟶ *Nano level*: Characterized by the interaction of free ends (fibrils), energy is bound, and the energetic status of the system is minimized, instability is possible.

Now, the general formation of new bonds at the nano level is investigated, and the entropy production is described.

$$\frac{dS_i}{dt} = X_B \cdot J_B \tag{263}$$

The thermodynamic force is generally described as X_B. It is assumed that the rate of structure formation φ_B, the solid content Γ, the number of free ends N, the proportion of free ends without interaction N_0, and thus the proportion of new bonds $(1 - N_0)$ influence it.

$$\frac{dS_i}{dt} = X_B \cdot J_B = \frac{X_B \cdot (\varphi_B \cdot \Gamma \cdot (1 - N_0) N)}{T} \tag{264}$$

A parameter ε is introduced to describe the distance to equilibrium. The disturbance of the steady state is observed.

$$\frac{\partial}{\partial t}(\delta^2 S) = \delta X_B \cdot \delta J_B = \frac{\partial X_B}{\partial \epsilon} \cdot \frac{\Gamma \cdot N}{T} \left((1 - N_0) \frac{\partial \varphi_B}{\partial \epsilon} - \varphi_B \frac{\partial N_0}{\partial \epsilon} \right) (\delta \epsilon)^2 \tag{265}$$

It is assumed that with increasing deviation from equilibrium, the thermodynamic force increases [46].

$$\frac{\partial X_B}{\partial \epsilon} > 0 \tag{266}$$

From the experimental investigations, we can assume that

$$\frac{\partial \varphi_B}{\partial \epsilon} < 0 \tag{267}$$

Thus, there is the possibility that an instability occurs in the process, and there can be conditions that allow for self-organization. This means the emergence of an unstable state on the *nanoscale* and the possible triggering of structure formation on the *mesoscale* [245].

10 Epilogue

In this book, I have attempted to approach the natural phenomenon of friction using the example of a stressed viscoelastic material. Considerations on the irreversible nature of the friction process are presented before the more traditional investigations into the behavior of lubricating greases.

Irreversibility and instability are no longer phenomena that receive negative evaluation and frequent neglect. Rather, they are the focal point and source of naturally occurring processes.

Departing from the user perspective and shifting to the perspective of the observed system is a paradigm shift in tribology and allows for a new understanding and initial answering of the question of *why*, as opposed to questions about the ongoing mechanisms.

The traditional description of the cause-effect chain with

Friction as the cause $\longrightarrow$ Wear as the effect

can now be changed to:

Friction as the cause $\longrightarrow$ for an **instability** and this **instability** is the cause $\longrightarrow$ for the **wear** to achieve **stability**

For the development of the tribological system, it seems important how far it is from thermodynamic equilibrium, regardless of the initial conditions.

In connection with space and time, situations can arise in which matter becomes active, organizes itself, and redefines energy dissipation.

Incorporating self-organization as a principle that determines nature into the understanding of friction raises completely new questions. Questions that extend beyond many tribological aspects into a *postmodern tribology.*

Hamburg in the year 2025, Prof. Dr.-Ing. E. Kuhn

11 Original Quotes

The original quotes by J.C.SCHMIDT from chapter 3.

In einer umfassenden Arbeit [3] schreibt J.C.SCHMIDT über die historische Entwicklung des Verständnisses von *Instabilität* „Das was sich als instabil zeigte, so nahm man an, sei lediglich eine Störung des Stabilen“.

SCHMIDT [3] drückt es noch prägnanter aus:„Nicht Stabilität, ... , sondern *Instabilität* gilt als Grundcharakter der Natur“.

An gleicher Stelle findet man eine terminologische Neudefinition. So schreibt [3]„Sofern die mathematischen Wissenschaften das *Instabile* in der Natur, Technik und Gesellschaft anerkennen und aussprechen, können sie als *nachmodern* bezeichnet werden “. „Der Begriff steht für eine sehr nüchterne Beschreibung der Veränderungen im Gefüge der Wissenschaften, ihrer Objekte, Methoden und Inhalte - von Mathematik über die Physik, Chemie, Biologie, Informatik und die Technik- und Ingenieurwissenschaften bis hin zu den Sozialwissenschaften.“

Here are the paragraphs from point 4.1 in German language with the original quotations from Kant, Schelling, and Goethe taken from the specified sources.

So beschreibt 1755 I. Kant [26] (Abb.(12)) Vorstellungen zur Bildung von Sternen und Planeten. Aus einem erfüllten Raum mit kleinsten Materieteilchen formen sich über Anziehung und Abstoßung letztendlich größere „Materieklumpen“ bis zu den Planeten. Dabei formuliert er Sätze wie „Die Elemente ... sind sich selbst Quelle des Lebens“ Oder noch markanter „... die Materie ... hat in ihrem einfachsten Zustande eine Bestrebung, sich durch eine natürliche Entwicklung zu einer voll kommeneren Verfassung zu bilden“. Bei der Beschreibung der Bewegungsbahnen der gebildeten Körper findet man: „In diesem Zustande..., da alle Theilchen ... durch die erlangten Schwungkräfte um den Zentralkörper laufen, ist der Streit und der Zusammenlauf der Elemente gehoben, und alles ist in dem Zustande der kleinsten Wechselwirkung.“

Eine Generation später schreibt SCHELLING in seinem Entwurf einer Naturphilosophie von 1799 [27] (Abb. (12)) in § VI,B „.. so liegt in der Stufenfolge (des Magnetismus, der Elektrizität und des chemischen Prozesses)(Anm. Klammern nicht im Originaltext), so wie sie auch am einzelnen Körper unterschieden werden kann, das Geheimnis der Produktion der *Natur aus sich selbst* “. Und weiter in §3, B3 „... denn diese (Anm. gemeint ist Natur) ist das Seyn oder die Produktivität selbst“. Und vorher „... Natur als Ganzes, das von sich selbst die Ursache zugleich und die Wirkung...“ Man findet unter § IV, A „... jenes Schweben der Natur zwischen Produktivität und Produkt wird also als eine allgemeine Duplicität der Prinzipien wodurch die Natur *in beständiger Tätigkeit* (nicht im Original kursiv) erhalten ... erscheinen müssen...“. Er schreibt in §III „... dass Bewegung nicht nur aus Bewegung, sondern selbst aus der Ruhe entspringe, dass also auch in der Ruhe der Natur Bewegung sei ...“. Und abschließend sei § II erwähnt mit „Denn alles Denken zuletzt auf ein Produzieren und Reproduzieren zurückkommt, so ist nichts unmögliches in den Gedanken, dass dieselbe Tätigkeit, durch welche die Natur in jedem Moment sich neu reproduziert...".

Nicht zuletzt finden wir im 18. bzw. frühen 19.Jhdt auch Überlegungen von J.W. VON GOETHE die in Relation zu unsrem heutigen Verständnis von Selbstorganisation verstanden werden können. So schreibt er „... und man darf daher ... eine unaufhaltsame fortschreitende Umbildung mit Recht annehmen "(zitiert in [28]).

Bibliography

[1] A. Knappwost. Wesen, Stellung und Aufgabe der Tribologie. *Schmierungstechnik und Tribologie*, 1973.

[2] I.Prigogine. *Vom Sein zum Werden (From Being to Becoming- Time and Complexity in Physical Sciences)*. Piper, 1988.

[3] J.C.Schmidt. *Das Andere der Natur*. S.Hirzel Verlag, 2015.

[4] Gesellschaft für Tribologie. Tribologie- Verschleiß, Reibung; Definitionen, Begriffe, Prüfung. *Arbeitsblatt7*, 2000.

[5] B.I. Kosjetzki. Grundlagen und Komplexverfahren für Problemlösungen in der Tribologie. *Schmierungstechnik*, 21(3), 1990.

[6] H. Czichos. Systemanalyse und Physik tribologischer Vorgänge. Teil 1. *Schmierungstechnik undTribologie*, 22(6), 1975.

[7] H.Czichos. Systemanalyse und Physik tribologischer Vorgänge. Teil 2. *Schmierungstechnik und Tribologie*, 23(6), 1976.

[8] G. Fleischer, G. Gröger, and H-Thum (unter Mitarb. von U. Winkelmann). Verschleiß und Zuverlässigkeit. *Verlag Technik*, 1980.

[9] G. Fleischer. Systembetrachtungen zur tribologie. *Wiss. Zeitschr. TH Otto von Guericke Magdeburg*, 14(5/6), 1970.

[10] G. Salomon. The structure of tribological systems. *Letter from Europe Lubr. Eng.*, 32(9), 1976.

[11] F. Bauer. *Tribologie prägnant und praxisrelevant*. Springer Viehweg, 2021.

[12] E. Kuhn. Besonderheiten des tribologischen Kontaktes bei Anwesenheit von Schmierfett. *Proceed. Schmierfett-Tagung, Fuchs Europe, Kiel*, 2008.

[13] E. Kuhn. Irreversible effects on grease structure due to friction energy. *6th International Conference on Mechanics and Material Design, Ponta Delgada, Prtugal.*, 2015.

[14] G. Fleischer. Probleme und Kriterien der Mischreibung. *Schmierungstechnik, Berlin*, 13(12), 1982.

[15] W. Gnilke. Der Einfluss der Druckabhängigkeit der Viskosität auf die Gleitlagerschmierung. *Schmierungstechnik*, 12(11), 1963.

[16] H. Umstätter. *Strukturmechanik.* Verlag von Theodor Steinkopff, 1948.

[17] E. Kuhn. Über die Triebkräfte des Schmierfettverschleißes. *Tribologie und Schmierungstechnik*, 65(6), 2017.

[18] E. Kuhn. Über die Unmöglichkeit des verschleißlosen tribologischen Prozesses. *Tribologie und Schmierungstechnik*, 54(6), 2007.

[19] E. Kuhn. Energy investigations of the rheological wear of lubricating greases. *Rheology*, 4, 1992.

[20] E. Kuhn. Grundlagen zur energetischen Berechnung des rheologischen Verschleißes von Schmierfetten. *Proceed. Tribol. Symp, ÖTG, Wien*, 1994.

[21] H. Czichos and M. Molgaard. Towards a general theory of tribological systems. *Wear*, 44, 1977.

[22] H.Czichos and G. Salomon. The application of systems thinking and systems analysis to tribology. *BAM-report Nr. 30*, 1974.

[23] H. Czichos and K.-H. Habig. *Tribologie-Handbuch. Tribometrie, Tribomaterialien, Tribotechnik.* Springer Fachmedien, 2015.

[24] I. Prigogine and I. Stengers. *Dialog mit der Natur. Neue Wege naturwissenschaftlichen Denkens.* R.Piper GmbH & Co. KG. München, 1990.

[25] E. I. Nkoyo. *Prigogines Theorie dissipativer Strukturen. Naturphilosophische und erkenntnistheoretische Betrachtungen.* Herbert Utz Verlag, 2016.

[26] I.Kant. *Allg. Naturgeschichte und Theorie des Himmels oder Versuch von der Verfassung und dem mechanischen Ursprunge des ganzen Weltgebäudes, nach Newtonschen Grundsätzen abgehandelt*, volume I. Verlag von Th. Knaur Nachf., um 1925.

[27] F.W.J. von Schelling. *Einleitung zu seinem Entwurf eines Systems der Naturphilosophie (1799).* Number ISBN/EAN: 978-3-74341-881-3. hanse, -.

[28] F. Schweitzer. *Goethes Morphologie-Konzept und die heutige Selbstorganisations-Theorie*, volume 3 in: Konzepte von Chaos und Selbstorganisation in der Geschichte der Wissenschaft of *Selbstorganisation. Jahrbuch für Komplexität in den Natur-Sozial- und Geisteswissenschaften, Bd.3.* Duncker | Humblot, Berlin, 1992.

[29] E.G. Assenova. On self-organization and selective transfer in tribological systems. *BALKANTRIB*, 2005.

[30] D.N. Garkunov, I.V. Kragelski, and A.A. Poljakov. *Selektive Übertragung in Reibstellen.* Moskau, Verlag Transport, 1969.

[31] B.I. Kostjetzki, I.G. Nosowski, and L.I. Berschadski. *Obcrflächenfestigkeit der Werkstoffe bei der Reibung.* Verlag Technika, 1976.

[32] B.I. Kostjetzki, M.E. Natanson, and L.I. Bershadski. *Chemisch-mechanische Prozesse bei der Grenzreibung.* Moskau Verlag Nauka, 1972.

[33] A.A. Poljakov. Eine Theorie der kohärenten Reibung - dissipative Strukturen bei der selektiven Übertragung. *Schmierungstechnik*, 21, 1990.

[34] B. E. Klamecki. Wear - an entropy production model. *Wear*, (58), 1980.

[35] B.E. Klamecki. Energy dissipation in sliding. *Wear*, (77), 1982.

[36] B.E. Klamecki. A thermodynamic model of friction. *Wear*, (63), 1980.

[37] G. Polzer and W. Ebeling. Die Reibung fester Körper, dissiptaive Strukturen und Selbstorganisation. *Schmierungstechnik, Berlin*, 17(1986)5.

[38] I.S. Gershman, E.I. Gershman, A.E. Mironov, G.S. Fox-Rabinovich, and S.C. Veldhuis. Application of self-organization phenomenon in the development of wear resistant materials - a review. *entropy*, 2016, 18,385.

[39] M. Nosonovsky. Entropy in tribology: in the search for applications. *entropy*, 2010,12.

[40] M.D. Bryant. Entropy and dissipative process of friction and wear. *FME Transaction*, 2009, 37.

[41] H. A. Abdel-Aal. *Thermodynamic modelling of wear.* Springer Science + buisness Media, 2011.

[42] M. Amiri and M.M. Khonsari. On the thermodynamics of friction and wear - a rieview. *entropy*, 2010,12.

[43] G.S. Fox-Rabinovich and G.E. Totten, editors. *Self-Organization during friction: Advanced surface-engineered materials and system design.* CRC Press, 2019.

[44] C. Viafara and A. Sinatora. Thermodynamic approaches in sliding wear: a review. *International Journal of Materials & Product Technology*, 38(1), 2010.

[45] M. Nosonovsky and B. Bushan. Surface self-organization: From wear to self-healing in tribological and technical surfaces. *Appl. Surf. Sc.*, (256), 2016.

[46] I.S. Gershman, E.I. Gershman, G.S. Fox-Rabinovich, and S.C. Veldhuis. Description of seizure process fo gas dynamic spray of metal powders from non-equilibrium thermodynamics standpoint. *entropy*, 2016,18,315.

[47] I.S. Gershman, N.A. Bushe, and A.E. Mironov. Self-organization and development of wearproof materials. *proceed. WTC, Washington*, 2005.

[48] M.D. Bryant, M.M. Khonsari, and F.F.Ling. On the thermodynamics of degradation. *Proc.R.Soc. A*, (464), 2008.

[49] M.D. Bryant and M.M. Khonsari. Application of degradation-entropy generation theorem to dry sliding friction and wear. *Proceed. STLE/ASME Intern Joint Conf.*, 2008 (IJTC2008-7107).

[50] F.F. Ling, M.D. Bryant, and K.L. Doelling. On irreversible thermodynamics for wear prediction. *Wear*, 253(11-12), 2002.

[51] M.D. Bryant. On constitutive relations for friction from thermodynamics and dynamics. *Transactions of ASME*, 138, 2016.

[52] J.A. Osara and M.D. Bryant. Performance and degradation characterization of electrochemical power sources using thermodynamics. *Electrochimica Acta*, (365), 2021.

[53] J.A. Osara. *The thermodynamics of degradation.* PhD thesis, The University of Texas at Austin, 2017.

[54] K.P. Lijesh and M.M. Khonsari. Characterization of abrasive wear using degradation coefficient. *Wear*, 450-451(15), 2020.

[55] M.A. Amiri and M.M. Khonsari. On the role of entropy generation in process involving fatigue. *entropy*, 14, 2011.

[56] K.P. Lijesh and M.M. Khonsari. On the modeling of adhesive wear with consideration of loading sequence. *Tribology Letters*, (2018) 66:105.

[57] Y.P. Kozyrev and E.B. Sedakova. Application of a thermodynamic model for analysis of wear resistance of materials. *Journal of Machinery, Manufacture and Reliability*, 37(1), 2008.

[58] E. Kuhn. The active role of lubricating grease as a tribological system - an energetic approach. *Proceed.TRIBOLOGY 2022-Tribology International Conference, 27.-29.4. Barcelona*, 2022.

[59] E. Kuhn. Analysis of a grease-lubricated contact from an energy point of view. *Int. J. Materials and Product Technology*, 38(1), 2010.

[60] S.R. De Groot and P. Mazur. *Grundlagen der Thermodynamik irreversibler Prozesse.* BI Mannheim/Zürich 1969.

[61] I.P. Basarow. *Thermodynamik.* Deutscher Verlag der Wissenschaften, Berlin, 1964.

[62] H.D. Baehr. *Thermodynamik.* Springer Verlag, 7. Aufl., 1989.

[63] G. Fleischer. Zum energetischen Niveau von Reibpaarungen. *Schmierungstechnik, Berlin*, 16(1), 1985.

[64] E. Kuhn. Correlation between system entropy and structural changes in lubricating greases. *lubricants*, 2015,3.

[65] E. Kuhn. Tribological stress of lubricating greases in the light of system entropy. *Lubricants*, 2016,4,37.

[66] R. Czarny. Einfluß der Thixotropie auf die rheologischen Eigenschaften der Schmierfette. *Tribologie und Schmierungstechnik*, 36(3), 1989.

[67] E. Kuhn. Aspects of self-organization of tribological stressed lubricating greases. *Lubricants*, 2020,8,28.

[68] G.E. Totten; G. Fox-Rabinovich, editor. *Self-Organization during friction: Advanced surface-engineered materials and system design.* CRC Press, 2007.

[69] G.Kluge and G. Neugebauer. *Grundlagen der Thermodynamik.* Deutscher Verlag der Wissenschaften, Berlin, 1976.

[70] *Thermodynamic theory of structure, stability and fluctuations.* Wiley Intersience, 1971.

[71] E. Kuhn and C. Balan. Experimental procedure for the evaluation of friction energy of lubricating greases. *Wear*, pages 237–240, 209(1997).

[72] N. Acar, J.M. Franco, and E.Kuhn. On the shear-induced structural degradation of lubricating greases and associated activation energy: An experimental rheological study. *Tribol. Int.*, 2020,144, 106105.

[73] G. Jetschke. *Mathematik der Selbstorganisation.* Vieweg-Verlag, Braunschweig/Wiesbaden, 1989.

[74] E. Kuhn. Application of thermodynamic concept for analysis of structural degradation of soap thickened lubricating greases. *Lubricants*, 2018, 6, 7.

[75] W. Holweger. Spectral density of lubricants in tribological contacts: towards a new sight of lubricants design for life. *Proceed. Tribology Coll., Ecole Polytechnique Federale de Lausanne*, 1999.

[76] W.J. Bartz. *Handbuch der Tribologie.* expert verlag, 1988.

[77] Hrsg. Richard Engelmann. *Leben der Griechen und Römer.* Berlin, Weidmannsche Buchhandlung, 1893.

[78] D. Todtenhaupt und A. Kurzweil. Holzteer in der Tribologie. *Tribologie und Schmierungstechnik*, 40(3), 1993.

[79] D. Klamann. *Schmierstoffe und verwandte Produkte.* Verlag Chemie, Weinheim, 1982.

[80] DIN. 51825. ., 1990.

[81] P.Lugt. *Grease lubrication in rolling bearings.* Tribology Series. John Wiley and Sons.Ltd, 2013.

[82] ASTM-D. 288-68. ., .

[83] Klüber Lubrication. Der Weg zur Schmierung. *München KG,9*, 1989.

[84] R. Czarny. Einfluss der Thixotropie auf die rheologischen Eigenschaften der Schmierfette. *Tribologie und Schmierungstechnik*, 36(3), 1989.

[85] W.H. Dresel. Grundlegende Aspekte zukunftsorienterter Schmierfette. *Tribologie und Schmierungstechnik*, 40(3), 1993.

[86] V.V. Sinitsin. The choice and application of plastic greases. *Chimija, Moskau*, 1974.

[87] H. Aström. *Grease in elastohydrodynamic Lubrication.* PhD thesis, Lulea University of Technology, 1993.

[88] O.Pigors. *Werkstoffe in der Tribotechnik.* Deutscher Verlag für Grundstoffindustrie, Leipzig/Stuttgart, 1993.

[89] W. Kohl. Schmierfette - Herstellung, Einsatz, Prüfung. *Tribologie und Schmierungstechnik*, 26(4), 1979.

[90] W. Holweger u.a. Stabilität und chaotisches Verhalten bei Schmierfetten. *Tribologie und Schmierungstechnik*, 37(2), 1990.

[91] J.Loges. *Auswirkungen der rheologischen Eigenschaften von Schmierfetten auf Zentralschmiersysteme, Gleitlager und Produktinformationen.* 1982.

[92] W.J. Bartz. *Schmierfette.* expert verlag, 2. Auflage, 2015.

[93] F. Wunsch. Leistngsfähigkeit von Schmierfetten auf Syntheseölbasis. *Tribologie und Schmierungstechnik*, 37(2), 1990.

[94] E.G. Pahland, G. Poll, and J.E. Franke. Ermittlung von Schmierfettkennwerten zum Reibungsverhalten und zur Schmierwirkungsdauer in schnelllaufenden Wälzlagern. *DGMK Projekt 379-1, Leibniz Universität Hannover*, 1979.

[95] *Lubricating grease guide,.* NLGI, second edition, Kansas City, 1987.

[96] FUCHS EUROPE SCHMIERSTOFFE GmbH. *Die Welt der Schmierstoffe.* FUCHS EUROPE SCHMIERSTOFFE GmbH, 2010.

[97] Y. Zhou, R. Bosman, and P.M. Lugt. A model for shear degradation of lithium soap grease at ambient temperature. *Tribology Transaction*, 12(21), 2016.

[98] M. Paszkowski. Identification of the thixotropiy of lithium greases. In W.J. Bartz, editor, *Tribology 2008 - Lubricants, Materials and Lubrication Engineering*, 2008.

[99] M. Grebe. Influence of mechanical, thermal, oxidative and catalytic processes on thickener structure and thus on the service life of rolling bearings. *Lubricants*, 10,77, 2022.

[100] R. Sanchez, C. Valencia, and J.M. Franco. Rheological and tribological characterization of a new acylated chitosian-based biodegradable lubricating grease: A comparative study with traditional lithium and calcium greases. *Tribology Transaction*, 57:3, 2014.

[101] N. Acar, E. Kuhn, J.M. Franco, D.E.P. Goncalves, and J.H.O. Seabra. Tribological investigation on the friction and wear behaviors of biogenic lubricating greases in steel-steel contact. *Applied Sciences*, 10(4), 2020.

[102] A.M. Borrero-Lpez, C. Valencia, and J.M. Franco. Rheology of lignin-based chemical oleogels prepared using diisocynate crosslinkers: Effect of diisocyanate and curingkinetics. *European Polymer Journal*, 89, 2017.

[103] G. Dobratz. Metallseifenschmierfette. *Tribologie und Schmierungstechnik*, 35(1), 1988.

[104] W. Mader. *Hinweise zur Anwendung von Schmierfetten.* Curt Vincent Verlag, 1979.

[105] F. Wunsch. Verwendung von schmierfetten mit angepassten eigenschaften. *Maschinenemarkt,*, 36(6), 1978.

[106] W.H. Dresel. Moderne schmierfette mit verlängerter lebensdauer. *Tribologie und Schmierungstechnik*, 36(6), 1989.

[107] W.J. Bartz. *Ökologische und ökonomische Aspekte bei Schmierfstoffen.* expert verlag, 2001.

[108] D. Kempkes. Einsatz biogener schmierstoffe in windkraftanalagen. *Tribologie und Schmierungstechnik*, 54(2007)3.

[109] FUCHS LUBRICANTS GERMANY GMBH. Biologisch schnell abbaubare schmierfette. *https://www.fuchs.com/de/de/produkte/schmierstoffe-und-services/schmierfette/biologisch-schnell-abbaubare-schmierfette/*, Stand 29.12.2023, 2023.

[110] M. Fiedler, R. Sanchez, E.Kuhn, and J.M. Franco. Influence of oilpolarity and material combination on the tribological response of greases formulated with biogredabable oils and bentonite and highly dispersed silica acid. *Lubrication Science*, 25, 2013.

[111] M. Fiedler. *Tribological and rheological investigation of the applicability of biogenic lubricating greases as separating medium in highly stressed friction systems.* PhD thesis, University of Huelva, 2014.

[112] R.Sanchez, J.M. Franco, M.A. Delgado, C.Valencia, and C. Gallegos. Effect of thermo-mechanical processing on the rheology of oleogels potentially applicable as biodegradable lubricating grease. *Chem. Eng. Res. Des.*, 86, 2008.

[113] R.Sanchez, J.M. Franco, M.A. Delgado, C.Valencia, and C. Gallegos. Development of new green lubricating greases formulations based on cellulosic derivates and castor oil. *Green. Chem.*, 83, 2009.

[114] R. Sanchez, J.M. Franco, E.Kuhn, and M. Fiedler. Tribological characterization of green lubricationg greases formulated with castor oil and different biogenic thickener agents: a comparative experimental study. *Industrial Lubrication and Tribology,*, 63, 2011.

[115] R. Sanchez, J.M. Franco, M.A. Delgado, C. Valencia, and C. Gallegos. Thermal and mechanical characterization of cellulosic derivates-based oleogels potentially applicable as bio- lubricating greases: Influence of ethyl cellulose molecular weight. *Carbohydr. Polym.*, 2011,83.

[116] R. Sanchez, G.B. Stringari, J.M. Franco, C. Valencia, and C. Gallegos. Use of chitin, chitosan and acylated derivates as thickener agents of vegetable oils for bio lubricant applications. *Carbohydr. Polym.*, 85, 2011.

[117] J.E.M. Alfonso, R. Yanez, C. Valencia, J.M. Franco, and M.J. Diaz. Optimization of the methylation conditions of Kraft cellulose pulp for its use as a thickener agent in biodegradable lubricating greases. *Ind. Eng. Chem. Res.*, 48, 2009.

[118] N. Nunez, J.E.M. Alfonso, C. Valencia, M.C. Sanchez, and J.M. Franco. Rheology of new green lubricating grease formulations containing cellulose pulp and its methylated derivates as thickener agents. *Ind. Crop. Prod.*, 37, 2012.

[119] E. Cortes-Trivino, C. Valencia, M.A. Delgado, and J.M. Franco. Modification of alkali lignin with poly(ethylene glycol) diglycidyl ether to be used as a thickener in bio-lubricant formulations. *Polymers*, 10, 670, 2018.

[120] E. Cortes-Trivino, C. Valencia, M.A. Delgado, and J.M. Franco. Rheology of epoxidized cellulose pulp gel-like dispersions in castor oil: Influence of epoxidation degree and epoxide chemical structure. *Carbohydrate Polymers*, 199, 2018.

[121] N. Acar, E. Kuhn, and J.M. Franco. Tribological and rheological characterization of new completely biogenic lubricating greases: a comparative experimental investigation. *Lubricants*, 6,45, 2018.

[122] N. Acar. *An eneretical approach to study the structural degradation of traditional and biogenic lubricating greases.* PhD thesis, University of Huelva, 2022.

[123] I. Slabka. Rheologische Charakterisierung biogener Schmierfette. Master's thesis, Hamburg University of Appl. Sc., 2017.

[124] J. Wilson. Three dimensional structure of grease thickener particles. *NLGI-Spokesman*, 27, 1964.

[125] S.F. Calhoun. Fundamental aspects of grease bleeding. *NLGI Spokesman*, 29, 1966.

[126] T. Mang and W. Dresel, editors. *Lubricants and lubrication.* Wiley-VCH, 2001.

[127] G. Spengler and F. Wunsch. Schmierung und Lagerung in der Feinwerktechnik. *VDI Verlag, Düsseldorf*, 1970.

[128] L. Ahme, E. Kuhn, and M.A. Delgado Canto. On the optical assessment of the structural degradation of rheologically stressed lubricating greases. *Tribology International*, 187, 2023, 108771.

[129] C. Roman, C. Valencia, and J.M. Franco. Afm and sem assessment of lubricating grease microstructures: influence of sample preparation protocol, frictional work condidtions and composition. *Tribology Lett*, 63(2):20, 2016.

[130] M.A. Delgado, M.C. Sanchez, C. Valencia, J.M. Franco, and C. Gallegos. Relationship among microstrukture, rheology and processing of a lithium lubricating grease. *Chemical Engineering Research and Design*, 83,9, 2005.

[131] J.W.H. Franke, J. Fritz, T. Koenig, and D. Merkanke and. Influence of tribolayer on rolling fatigue performed on an fe8 test rig - a follow ou. *Lubricants*, 2023,11(3), 123.

[132] G. Bayer, A. Bartschat, S. Wandel, S. Baust, and G.Poll. Experimental investigation on wear in oscillating grease-lubricated rolling element bearings of different size and type. *Lubricants*, 2013, 11(3), 120.

[133] F. Ebert. *Strömung Nicht-Newtonscher Medien.* Vieweg-Verlag, Braunschweig/Wiesbaden, 1980.

[134] M. Reiner. *Rheologie in elementarer Dartsellung.* Carl Hanser Verlag, München, 1968.

[135] G. Böhme. *Strömungsmechanik Nicht-Newtonscher Fluide.* Teubner Verlag, Stuttgart, 1981.

[136] R. Spaltofski. *Ein Beitrag zur Beurteilung der Trageigenschaften und des Reibungsverhaltens von Nicht-Newtonschen Flüssigkeiten in stationär und dynamisch bewegten Gleitlagern.* PhD thesis, Universität Kaiserslautern, 1988.

[137] W.-M. Kulicke. Fließverhalten von Stoffen und Stoffgemischen. 1986.

[138] K. Weissenberg. *Nature*, 159, 1947.

[139] DIN. 13342. 06.76.

[140] Peterfi. *Arch. f. Entwicklungsmechanik d. Organis.*, 112,660, 1927.

[141] A. Szegvari and E. Schalek. *Koll.-Zeitschr.*, 32, 318, 1923.

[142] A. Szegvari and E.Schalek. *Koll.-Zeitschr.*, 33,326, 1923.

[143] H. Freundlich. Über Thixotropie. *Koll.-Zeitschr.*, 46, 1928.

[144] W. Bohl. *Technische Strömungslehre.* Vogel Verlag, Würzburg, 9.aufl. edition, 1991.

[145] H. Freundlich and F. Juliusburger. Thixotropy, influenced by the orientation of anisometric particles in sols and suspensions. *The Sir William Ramsay Laboratories of Phys. and Inorg. Chem, Univ. College. London*, 1935.

[146] A. Dunker. Untersuchungen zur Rheologie strukturviskoser Schmierstoffe. Master's thesis, Hamburg University of Appl. Sc., 1997.

[147] R. Czarny. Temperaturabhängigkeit einiger rheologischer Parameter von Schmierfetten. *Tribologie und Schmierungstechnik*, 37(3), 1990.

[148] H. Åström and E. Höglund. Rheological properties of six greases and their two base oils. *Research report, Luleå Univ.*, 1990.

[149] J. Loges. Auswirkungen der rheologischen Eigenschaften von Schmierfetten auf Zentralschmiersysteme, Gleitlager und Produktinformationen. *Habil.schrift, Otto von Guericke Univ. Magdeburg*, 1982.

[150] W.H. Bauer, A.P. Finkelstein, and S.E. Wiberley. Shear behaviour of elastohydrodynamic oil films. *Proc. R.Soc. London,*, Serie A(356), 1977.

[151] S. Bair. The high-pressure rheology of soaped-thicked grease. *Trib. Transact.*, 37(3), 1994.

[152] S. Yasutomi, S. Bair, and W.O. Winer. An application of a free volume model to lubricant rheology. *ASME Journal of Tribology*, 106(2), 1984.

[153] W.H. Herschel and R. Bulkley. *Kolloid-Zeitschr.*, 39, 1926.

[154] D. Klamann, L. Endom, R. Rost, and A. Haak. Zur rheologie der schmierfette. *Erdöl und Kohle, Erdgas, Petrochemie*, 20, 1967.

[155] J.M. Palacios and M.P. Palacios. Rheological properties of grease in ehd contact. *Tribology International*, 17, 1984.

[156] K. Stanulov, P. Dewensky, and J. Kamberova. Der einfluss einiger polymere auf die rheologischen eigenschaften von lithium-schmierfetten. *Hochschule für chem. Techn., Sofia, Bulgarien.*

[157] J. Wilson and G.H. Smith. Calculating pressure flow relations for lubricating greases. *Industrial and Engineering Chemistry*, 42(4), 1949.

[158] G.A. Plint and A.F. Alliston-Greiner. A new grease viscosimeter: A study of the influence of shear on the properties og greases. *NLGI Spokesman*, 56(2), 1992.

[159] K. Spiegel, J. Fricke, K.-R. Meis, and F. Sonntag. Die Fließeigenschaften von Schmierfetten in Abhängigkeit von Beanspruchung, Beanspruchungsdauer, und Temperatur. *Proceed. Tribology 2000 Intern. Coll. Esslingen*, 1992.

[160] E. Kuhn. Energy input to attain a minimum viscosity. *Rheology*, 7, 1995.

[161] E. Dewald. Mehr rheologie in die tribologie (interview). *Rheology*, 7, 1995.

[162] E. Kuhn. Tribological interpretation of rheological parameter. *Rheology*, 6, 1994.

[163] E. Kuhn. Energy interpretation of thixotropic effects. *Wear*, 142, 1991.

[164] E. Kuhn. Energetical investigations of the tribology of lubricating greases. *CEC and SAE Meeting*, 2000, Paris.

[165] C. Balan and J.M. Franco. *Influence of the geometry on the rotational rheometry of lubricating greases.* ELGI, 2000.

[166] P. Lugt. *Grease lubrication in rolling bearings.* Wiley and Sons, Ltd. Tribology Series, 2013.

[167] T. Mezger. *Das Rheologie-Handbuch.* Vincentz, 2000.

[168] M.A. Delgado, J.M. Franco, and E.Kuhn. Effect of rheological behaviour of lithium greases on the friction process. *Industrial Lubrication and Tribology*, 60(1), 2008.

[169] T. Litters and B. Koch. Einfluss der basisölpolarität auf das viskoelastische verhalten von schmierfetten. *Proc. GfT-Tagung, Göttingen*, 2008.

[170] A. Tross. *Über das Wesen und den Mechanismus der Festigkeit.* Eigenverlag, Zell am See, 1966.

[171] M.Dzimko. *Analytische Bestimmung der Mischreibung metallischer Gleitpaarungen auf energetischer Grundlage.* PhD thesis, Otto von Guericke University Magdeburg, 1982.

[172] E. Kuhn. Reibungsverhalten beim KontaktmodellKugel-Kugel. *Schmierungstechnik, Berlin*, 16(12), 1985.

[173] F. Rühle. Erxperimentelle ermittlung der eigenschaften oberflächennaher stoffbereiche bei unterschiedlichen fertigungsverfahren. *Proceed.3. Arnold Tross Kolloquium, Hamburg*, 2007.

[174] U. Winkelmann. *Grundlagen zur energetischen Bestimmung von Reibungskenngrößen bei Festkörperreibung metallischer Gleitpaarungen.* PhD thesis, Otto von Guericke University Magdeburg, 1981.

[175] W. Thomas. *Reibung und Verschleiß an Reibpaarungen mit Gleitbewegung beim Wirken abrasiver Teilchen.* PhD thesis, Otto von Guericke University Magdeburg, 1979.

[176] I.P. Kragelski, M.N. Dobicin, and V.S. Kombalov. *Reibung und Verschleiß.* Verlag Technik, 1982.

[177] B. Patzelt. Simulation einer stoßartigen beanspruchung mit dem pendelfurcher. *Diss. Otto von Guericke Universität, Magdeburg*, 1995.

[178] E. Kuhn. Verschleißuntersuchungen am Kontaktmodel Kugel-Kugel. *Schmierungstechnik, Berlin*, 17(9), 1986.

[179] K. Kuhn and E. Kuhn. Untersuchungen zur kritischen Anzahl der Kontaktierungen unter Anwendung des kumulativen stochastischen Prozesses. *Wiss. Zeitschr. der Otto von Guericke Universität Magdeburg*, 34(2), 1990.

[180] K. Kuhn and E. Kuhn. Investigations of the critical number of contacts with a mathematical model. *Proceed. EuroTrib, Budapest*, Vol.1, 1994.

[181] G. Vogelpohl. Reibung und Verschleiß in wirklichen Maschinen zur Beurteilung von Forschungsergebnissen. *Forsch.-Inbg. Wes.*, 26, 1960.

[182] J. Sadowski. Thermodynamisches Grundgesetz des tribologischen Verschleißes. *Tribologie und Schmierungstechnik*, 37(2), 1990.

[183] J. Sadowski. Die Verschleißzahl. *Tribologie und Schmierungstechnik*, 54(4), 07.

[184] G. Fleischer. Zur Energetik der Reibung. *Wiss. Zeitschr. der Otto von Guericke Universität Magdeburg*, 1990.

[185] M. Planck. Über die Begründung des zweiten Hauptsatzes der Thermodynamik. *Sitzungsbericht, Berl. Akad.Physk.-Mathem. Klasse*, pages 453–463, 1926.

[186] C.-F. von Weizsäcker. Die philosophische Interpretation der moderenen Physik. *Nova Acta Leopoldina, Halle Saale*, 207(37,2), 1975.

[187] E. Kuhn. Rheological wear as a result of the tribological stress in a hard - hard contact. *Proceed. 3rd Arnold Tross Colloquium, Hamburg,*, 2006.

[188] M.A. Delgado, C. Gallegos, J.M. Franco, and E. Kuhn. An experimental study of relationship between the time dependent rheological response and the friction process of lithium greases. *Proseed. 3rd Arnold Tross Colloquium, Hamburg*, 2007.

[189] R. Czarny. Effect of changes in grease structure on sliding friction. *Industrial Lubrication and Tribology,*, 47(1), 1995.

[190] E. Kuhn. Tribological analysis of a grease lubricated contact. *STLE-Meeting, New York*, 2003.

[191] H. Müller and H. Daum. Beitrag zur Klärung des Zusammenhangs zwischen Werkstoffbeanspruchung und Verschleiß beim Gleiten metallischer Reibpaarungen. *Diss. Otto von Guericke Universität*, 1974.

[192] E. Kuhn. Einfluss der rheologischen Eigenschaften auf den tribologiachen Prozess. *Schmierungstechnik, Berlin*, 21(4), 1990.

[193] E. Kuhn and W. Holweger. Einfluss der Schmierfettstruktur auf das tribologische Verhalten. *Proc. 12th Intern. Coll. Tribology, Esslingen, Ed. J.W. Bartz*, 2000.

[194] M.A. Delgado, J.M. Franco, and C. Valencia. Combined oxidative - shear re- sistance of castor oil - based greases. *Proceed. 3rd Arnold Tross Colloquium, Hamburg*, 2006.

[195] E. Kuhn. A strain-rate-diagram for lubricating greases. *Eurogrease*, 8/9, 2001.

[196] E. Kuhn. Tribological characterization of lubricating greases by using energetical parameter. *25th School of Tribology, Univ. of Techn. Wroclaw*, 2002.

[197] E. Kuhn and T. Schmidt. Cohesion behaviour of lubricating greases with a new pendulum tribometer. *Eurogrease*, 11/12, 2002.

[198] A. Gurt and M. Khonsari. Comparison of rheological methods to measure grease degradation. *Lubricants*, 11,468, 2023.

[199] W. Gnilke. Der Einfluss der Druckabhängigkeit der Viskosität auf die Gleitlagerschmierung. *Maschinenbautechnik, Beilage Schmierungstechnik*, 12(11), 1963.

[200] U. Winkelmann. Analyse der tribologischen Beanspruchung beim Wälzen. *Schmierungstechnik, Berlin*, 17(9), 1986.

[201] J. Hadler and G. Fleischer. Radialgleitlager undet Bedingungen der Flüssigkeits- und Mischreibung. *Antriebstechnik*, 34(2), 1995.

[202] X. Liebl and E. Vamos. Anregungen zum Aufbau einer Kolloidtheorie plastischer Schmierstoffe. *Tribologie und Schmierungstechnik,*, 15, 1968.

[203] B.W. Hotten. Optical studies of lubricating greases structure. *NLGI Spokesman*, 12, 1955.

[204] J.-S. Merieux, S. Hurley, A.A. Lubrecht, and P.M. Cann. Shear-degradation of grease and base oil availability in strved ehl lubrication. *Proceed. 27th Leeds-Lyon Conference*, 1999.

[205] E. Kuhn. *Zur Tribologie der Schmierfette. Eine energetische Betrachtungsweise des Reibungs- und Verschleißprozesses.* expert verlag, 2nd edition, 2017.

[206] E. Kuhn. Effect of friction energy for a grease lubricated contact. *13th NordTrib, Tampere*, 2008.

[207] M.C. Sanchez, J.M. Franco, C. Valencia, C. Gallegos, F. Uequiola, and R. Urchegui. Atomic force microscopy and the thermorehological characterization of lubricating greases. *Tribol. Letters*, 41:463–470, 2011.

[208] M. Paszkowski and S. Olsztynska-Janus. Grease thixotropy: evaluation of microstructure change due to shear and relaxation. *Industrial Lubrication and Tribology*, 66(2), 2014.

[209] M. Paszkowski and Sylwia Osztynska-Janus. Research on the kinetics of lithium grease thixotropic microstructure reconstruction. *proceed. WTC, Torino*, 2013.

[210] A. Rezasoltani. On the degradation of lubricating grease. *PhD-Thesis, Louisiana State University*, 2016.

[211] M. M. Khonsari, K. P. Lijesh, R. A. Miller, and R. Shah. Evaluating grease degradation through contact angle approach. *Lubricants*, 9(11), 2021.

[212] K.P. Lijesh and M.M. Khonsari. On the assessment of mechanical degradation of grease using entropy generation rate. *Tribology Letters*, 67:50, 2019.

[213] A. Gurt and M. Khonsari. The use of entropy in modeling the mechanical degradation of grease. *Lubricants*, 7,82, 2019.

[214] J.A. Osara and M.D. Bryant. Thermodynamics of grease degradation. *Tribology International*, 137, 2019.

[215] T. Rieling. Low temperature rjeometer tests of lubricating greases. *Poster exhibition, Proc. 10th Arnold Tross Colloquium Hamburg*, 2014.

[216] L. Ahme, E. Kuhn, and M. Delgado. Experimental study on the expended energy on structural degradation of lubricating greases. *Tribol. Letters*, 70:81, 2022.

[217] l. Ahme, E. Kuhn, and M.A. Delgado. An approach of the internal friction-dependent temperature changes for conventional and pure biogenic lubricating greases. *Friction*, https://doi.org/10.1007/s40544-023-0818-7, 2013.

[218] H. Hertz. Über die Berührung fester elastischer Körper. *J. f. die reine und angewandte Math. Bd.92*, 1881.

[219] B. Tower. First report on friction experiments (friction of lubricated bearings). *Proc.Instn. Mech. Eng. Nov.*, 1883.

[220] O. Reynold. On the theory of lubrication and its application to mr. beauchamo towers experiments, including an experimental determination of the viscosity of olive oil. *Phil. Trans. R. Soc. Bd.177*, 1886.

[221] A. v. Mohrenstein-Ertel. Die Berechnung der hydrodynamischen Schmierung gekrümmter Oberflächen unter hoher Belastung und Relativbewegung. *VDI-Fortschrittsberichte, Reihe 1, Nr. 115 (Übersetzung der russischen Diss. 1944)*, 1984.

[222] M.J.W. Schouten and H.J. v. Leeuwen. Die Elastohydrodynamik: Geschichte und Neuentwicklungen. *VDI-Berichte Nr. 1207*, 1995.

[223] P.M. Lugt, M.T. v. Zoelen, and C.H. Venner. *Film thickness theory for single contacts (Chapter 9 in Grease Lubrication in Rolling Bearing,P.M. Lugt).* John Wiley and Sons, Ltd., 2013.

[224] H. Åström, J. O. Östensen, and E. Höglund. Lubricating grease replenoshment in an elastohydrodynamic point contact. *J. Tribol.*, 115, 1993.

[225] P. Cann and H. Spikes. Thin film optical interferometry in the study of grease lubrication in a rolling contact. *Acta Tribologica*, 2, 1994.

[226] P. Cann. Grease lubricant films in rolling contact. *ELGI-Conf. 1997.*

[227] E. Höglund. Ehl and the use of image analysis. *Wear*, 179, 1994.

[228] S. Hurley and P. Cann. Grease composition and film thicjness in rolling contacts. *NLGI Spokesman*, 62, 1999.

[229] H. Åström. *Grease in elastohydrodynamic lubrication.* PhD thesis, LuleåUniversity, 1993.

[230] X. Li, F. Guo, G. Poll, and Y. Fei andP. Yang. Grease film evolution in rolling elastohydrodynamic lubrication contacts. *Friction*, 9,1, 2020.

[231] D. Fischer, G. Jacobs, A. Stratmann, and G. Burghardt. Effect of base oil type in grease composition on the lubricating film formation in ehd contacts. *Lubricants*, 6,32, 2018.

[232] D.E.P. Goncalves, A. V. Campo, and J.H.O. Seabra. An experimental study on starved grease lubricated contacts. *Lubricants*, 6,82, 2018.

[233] M. Löhr, E. Kuhn, T. Rieling, Ö. Ölcen, and D. Rajabi. Acoustic emission of lubricating greases under shear stress - rheoacoustics (in german). *18. Kolloquium Schallemission, Wetzlar*, 2011.

[234] M.Löhr, E. Kuhn, and T. Rieling. Acoustic emission of lubrication greases in rheometer. *DGzfP-Schallemissionskolloquium, Bad Schandau*, 2009.

[235] Ö. Ölcen and D. Rajabi. Acoustic emission measurement on biogenic model greases under rheological load (in german). Master's thesis, Hamburg Univ. of Appl. Sc., 2011.

[236] Ö. Ölcen. Investigation of biogenic lubricating greases with regard to the influence of polarity with energetic interpretation (in german. Master's thesis, Hamburg Universitry of Appl. Sc., 2012.

[237] F. Kaiser, M. Detjens, and F. Kelmendi. Development and application of a procedure to investigate the tribological behavior of biogenic greases (in german). Technical report, Hamburg University of Appl. Sc., 2016.

[238] M. Fiedler, R. Sanchez, C. Valencia, C.S. Leupold, E. Kuhn, and J. M. Franco. Influence of base oil poaritry on the transient shear flow of biogradable lubricating greases. *Lubricants*, 3, 2015.

[239] H. Freundlich and D. Krüger. Anomalous diffusion in true solution. *Transactions of the Faraday Society*, 31, 1935.

[240] H. Freundlich. Colloidal structures in biology. *University College, London*, 1937.

[241] H. Freundlich and F. Juliusburger. The plasticity of powdered slate from solnhofen and the thixotropic behaviour of its suspensions. *Transactions of the Faraday Society*, 30, 1934.

[242] H. Freundlich. The structure and formation of colloidal particles. *Transactions of the Faraday Society*, 23, 1927.

[243] H. Freundlich and W. Rawitzer. Über die thixotropie des konzentrierten eisenoxydsols. *Kolloidchemische Beihefte Band XXV*, (5-8), 1927.

[244] F. Sigmund. Einfluss von Temperatur und Feststoffanteil auf das thixotrope Verhalten ausgewählter Modellfette: Eine experimentelle Untersuchung. Technical report, HAW Hamburg, TREC, 2023.

[245] E. Kuhn. Some considerations on the formation of dissipative structures in grease films. *Lubricants*, 13(86), 2025.

List of Figures

Index of Persons

Index